COSMOLOGY AND POLITICAL CULTURE IN EARLY CHINA

Cosmology and a unified empire have long been considered the two most enduring structures of Chinese civilization. The role of cosmology in the formation of China's early empires is a vital question for historians of China and one with great relevance to the definition of "Chineseness" today. This book offers a radical reinterpretation of the formative stages of Chinese culture and history, tracing the central role played by cosmology in the development of China's early empires. Aihe Wang unveils the dynamic interaction between these two legacies – the cultural and the political – in the historical process.

Wang examines the transformation of Chinese cosmology between two political eras – from the hegemonic states of the Bronze Age (the Shang and Western Zhou, ca. 1700–771 B.C.) to the unified empires of the Iron Age (Qin and Han, 221 B.C.–220 A.D.). Challenging the prevailing view of cosmology as a quintessential, unchanging, homogenous structure of Chinese culture, she demonstrates how cosmology was constructive to power while being at the same time constantly transformed by the political process. The ruling clans of the Bronze Age drew legitimacy through a cosmological system known as *Sifang* (the Four Quarters), in which the king and his ancestral line were believed to be the conduit of divine authority. Wang illustrates how beginning in about 400 B.C., the shift to *Wuxing* (commonly known as the Five Elements, in which the cosmic energies of wood, fire, earth, metal, and water constantly interact) both paved the way for and was subsequently refined by the politics of a unified, imperial order. Engaging social theory as well as philosophical, historical, and anthropological approaches, the author offers a model of dynamic and multifaceted political discourse as an alternative to the prevailing, more narrowly conceived theories of culture and power.

Aihe Wang is an assistant professor in the department of history at Purdue University. She grew up in Beijing and earned an M.A. from the Chinese Academy of Social Sciences in 1986. In 1995 she received a Ph.D. in social anthropology and East Asian languages and civilizations from Harvard University, supported by a Harvard Yen-ching scholarship. She has published in *East Asian Archaeology*, *Daojia Wenhua Yanjiu* [The Study of Daoist Culture], and various anthologies on Chinese history and culture.

Cambridge Studies in Chinese History, Literature and Institutions

Victor H. Mair Tunhuang Popular Narratives
Ira E. Kasoff The Thought of Chang Tsai
Chih-P'ing Chou Yüan Hung-tao and the Kung-an School
Arthur Waldron The Great Wall of China: From History to Myth
Hugh R. Clark Community, Trade, and Networks: Southern Fujian Province from the Third to the Thirteenth Centuries
Denis Twitchett The Writing of Official History Under the T'ang
J. D. Schmidt Stone Lake: The Poetry of Fang Chengda
Brian E. McKnight Law and Order in Sung China
Jo-Shui Chen Liu Tsung-yüan and Intellectual Change in T'ang China, 773–819
David Pong Shen Pao-chen and China's Modernization in the Nineteenth Century
J. D. Schmidt Within the Human Realm: The Poetry of Huang Zunxian, 1848–1905
Arthur Waldron From War to Nationalism: China's Turning Point, 1924–1925
Chin-Shing Huang Philosophy, Philology, and Politics in Eighteenth-Century China: Li Fu and the Lu-Wang School under the Ch'ing
Glen Dudbridge Religious Experience and Lay Society in T'ang China: A Reading of Tai Fu's 'Kuang-i chi'
Eva Shan Chou Reconsidering Tu Fu: Literary Greatness and Cultural Context
Frederic Wakeman Jr. The Shanghai Badlands: Wartime Terrorism and Urban Crime, 1937–1941
Sarah A. Queen From Chronicle to Canon: The Hermeneutics of the Spring and Autumn Annals according to Tung Chung-shu
J. Y. Wong Deadly Dreams: Opium, Imperialism, and the *Arrow* War (1856–1860) in China
Norman Kutcher Mourning in Late Imperial China

Cosmology and Political Culture in Early China

Aihe Wang
Purdue University

CAMBRIDGE
UNIVERSITY PRESS

CAMBRIDGE UNIVERSITY PRESS
Cambridge, New York, Melbourne, Madrid, Cape Town, Singapore, São Paulo

Cambridge University Press
The Edinburgh Building, Cambridge CB2 2RU, UK

Published in the United States of America by Cambridge University Press, New York

www.cambridge.org
Information on this title: www.cambridge.org/9780521624206

First published 2000
This digitally printed first paperback version 2006

A catalogue record for this publication is available from the British Library

Library of Congress Cataloguing in Publication data

Wang, Aihe, 1954–
Cosmology and political culture in early China / Aihe Wang.
p. cm. – (Cambridge studies in Chinese history, literature and institutions)
Includes bibliographical references.
ISBN 0-521-62420-7 hardback
1. Cosmology, Chinese. 2. Political culture – China. 3. China – Politics and government – To 221 B.C. 4. China – Politics and government – 221 B.C. to 220 A.D. I. Title. II. Series.
BD518. C5W29 1999
181′. 11 – dc21 99–24301

ISBN-13 978-0-521-62420-6 hardback
ISBN-10 0-521-62420-7 hardback

ISBN-13 978-0-521-02749-6 paperback
ISBN-10 0-521-02749-7 paperback

To my teacher,
Kwang-chih Chang

Contents

Illustrations

Figures

Maps

Tables

Acknowledgments

This book is the fruit of more than a decade of inspiration and guidance from my teachers, Professors Kwang-chih Chang, Michael Loewe, Sally F. Moore, and Stanley J. Tambiah. Many other scholars have also supported this project over the years. Lothar von Falkenhausen has encouraged me at every stage and provided invaluable commentary on several incarnations of the manuscript. Nathan Sivin and David Keightley commented extensively on earlier versions of the text and generously shared their own works in progress. Edward Shaughnessy has been most kind, furthering scholarly dialogue as well as commenting on Zhou materials. Constance Cook scrutinized Chapter 2 and shared her expertise working with Zhou sources; Li Feng read revised portions of that chapter. Robin Yates familiarized me with many newly unearthed documents, and Wu Hung helped with architectural and visual materials. Early in my research, I was also inspired and assisted in various ways by Benjamin Schwartz, Hsu Cho-yun, Peter Bol, Stephen Owen, and Tu Wei-ming.

For my work in anthropology, I thank Michael Herzfeld, James Ferguson, Lissa Malkki, Rubie Watson, and James Watson for their inspiration and encouragement. Angela Zito has been especially supportive. While opinions differ among the many scholars I have mentioned and the many others whose work has made this book possible, I am grateful to all of them for helping me across a vast temporal and intellectual terrain.

For the production stages of the book, I am indebted to Denis Twitchett, the series editor for Cambridge Studies in Chinese History, Literature and Institutions, for his valuable suggestions for revising the manuscript; to Wann Ai-jen for editing the Chinese text and bibliography; and to Russell Hahn for his fine editing of the final copy. The book and the dissertation from which it grew received financial support from the Harvard Yen-ching Institute, the Harvard Graduate School of Arts and Sciences, and the Purdue Research Foundation. Research at Harvard was facilitated by many people at Yen-ching, Widener, Tozzer, and Rubell Libraries. I also thank the Fairbank Center for East Asian Studies at Harvard and the Center for Chinese Studies at the University of Chicago for offering me associate memberships and access to their facilities.

Finally, I would like to thank my husband, Greg Thomas, and my family for sustaining me throughout the process of research and writing.

1
Introduction

Cosmology and Political Culture

This work examines the transformation of Chinese cosmology between two political ages – from the hegemonic states of the Bronze Age (Shang and Western Zhou, ca. 1700–771 B.C.) to the unified empires of the Iron Age (Qin and Han, 221 B.C.–220 A.D.) This historical transition produced two enduring traditions of Chinese civilization: the cultural heritage of a cosmology that has been seen as a "primordial and quintessential expression of the 'Chinese Mind'" or the "Chinese 'structure of thought'";[1] and the political heritage of a unified empire that has been considered the ideal model of Chinese government ever since. The task of this book is to unveil the interrelations and mutual production of these two heritages – the cultural and the political – in the historical process.

The role of cosmology in the formation of China's early empires is a crucial question in Chinese history, one with great relevance to defining "Chineseness" today. This is because cosmology and the unified empire have been seen as the two most enduring structures of Chinese civilization. Two thousand years of official histories have repeatedly told the story of their eternal validity, transcending time and events, so that this unchanging order has become an unquestionable truth. Today, cosmology and a unified empire still serve as resources for forging China's national identity. Revived by some, cursed by others, traditional cosmology is used to represent a cultural identity that is authentically Chinese, and a unified empire continues to be held by most Chinese as the only justified form of government for China.[2] By questioning the

1 This phrasing is borrowed from Schwartz, 1985, p. 351.

2 The search for identity at national and personal levels in modern China is discussed in Dittmer and Kim, eds., 1993, and Tu Wei-ming, ed., 1991. The rebel Chinese voices on this issue are best represented in Barmé and Jaivin, eds., 1992.

social production of these two enduring structures of Chinese civilization, this study seeks to demonstrate how, beneath their unitary and recurring patterns, cosmology as a realm of the cultural and empire as a realm of the political were formed by a common dialectical process of mutual production and transformation in early China.

Chinese cosmology has been characterized as "correlative." Cosmologies, using anthropologist Stanley Tambiah's definition, are "frameworks of concepts and relations which treat the universe or cosmos as an ordered system, describing it in terms of space, time, matter, and motion, and peopling it with gods, humans, animals, spirits, demons, and the like."[3] Chinese cosmology, as such a framework of conceptions and relations, is an immense system of correlation-building, based on interlaced pairs (correlated to *Yin-Yang* 陰陽), fours (correlated to the four directions), fives (correlated to the Five Phases or *Wuxing* 五行), eights (correlated to the Eight Trigrams), and so on. Such a correlative cosmology is an orderly system of correspondence among various domains of reality in the universe, correlating categories of the human world, such as the human body, behavior, morality, the sociopolitical order, and historical changes, with categories of the cosmos, including time, space, the heavenly bodies, seasonal movement, and natural phenomena. Schwartz has found that Chinese correlative cosmology resembles what Lévi-Strauss describes as the "science of the concrete" – "a kind of anthropocosmology in which entities, processes, and classes of phenomena found in nature correspond to or 'go together with' various entities, processes, and classes of phenomena in the human world."[4]

A mode of thinking that has appeared in most civilizations,[5] correlative cosmology nevertheless has different functions and meanings in different cultures and historical environments. In China, its first cultural-political manifestation occurred during the formative stage of China's early empires in the last four centuries B.C. It was during this political transition that correlative cosmology became a common discourse by means of which competing social forces argued with one another, contested over the order of the new empire, and prescribed social practices in daily life. As such a common discourse, Chinese cosmology became a prevalent expression of political culture that was essential to the formation of the imperial order of early China, which continued to influence imperial history for two thousand years.

3 Tambiah, 1985, p. 3. 4 Schwartz, 1985, p. 351.

5 A. C. Graham has powerfully demonstrated that far from being an "exotic" mode of thinking uniquely Chinese, Chinese correlative cosmology-building is merely an example of the "correlative" thinking used by everyone, which underlies the operations of language itself. Graham, 1989, p. 320.

The development of the core of Chinese correlative cosmology – the system called *Wuxing* – best illustrates how cosmology and the imperial formation were mutually productive. *Wuxing* is a cosmology symbolized by the five material elements – Wood, Fire, Earth, Metal, and Water. It is a system of classification that became predominant over other systems, synthesizing and standardizing the other systems through these five categories. Yet more important than its role as a means of classification, *Wuxing* is a cosmology about interaction and change. The five cosmic energies exist in constant interaction, conquering and generating one another in circular sequence. As the core of correlative cosmology, *Wuxing* correlated events and actions with the ceaseless cosmic movements of the five interactive phases, serving to explain events in the human world and to dictate human actions.

The proper translation of the term *Wuxing* 五行 has long been debated by scholars. The traditional translation is "Five Elements," a term most convenient for comparative studies of Chinese thought and thought in other civilizations.[6] Yet "elements" does not fully represent the Chinese term *Wuxing*, which literally means five "goings," "conducts," or "doings," nor does it convey the basic nature of *Wuxing* as a cosmology of interaction and change. Many scholars have proposed alternatives, including five forces, agents, entities, activities, or stages of change.[7] Of these, "Five Phases" has acquired a wide acceptance among specialists.[8] But some scholars have recently challenged "Five Phases,"[9] among them A. C. Graham, who uses "Five Processes" for the pre-Han period.[10]

This difficulty in translation derives primarily from *Wuxing* cosmology itself, from its fluidity and diversity in function and meaning. As I shall demonstrate throughout this study, its meaning varied not only in different historical periods, but also in its different applications by diverse factions in the same society. *Wuxing* is not simply a set of concepts, a school of philosophy, a mode of thinking, or a commonly agreed-upon representation; instead, it is a cultural phenomenon that changes through history, a discourse for political argument and power struggle, and above all, an art of action in a world of conflict and change. Political actors used *Wuxing* cosmology in arguing about imperial sovereignty,

6 "Five Elements" has been used by scholars such as Marcel Granet and Derk Bodde.
7 See the list in Kunst, 1977.
8 John S. Major has strongly argued for using "Five Phases." See Major, 1976; 1977.
9 Michael Friedrich and Michael Lackner suggest restoring the term "elements." See Friedrich and Lackner, 1983–5. Bodde agrees with this suggestion and favors "elements" over "phases," saying that the latter does not convey the dynamism of *Wuxing* in ceaseless interaction. See Bodde, 1991, p. 101.
10 Graham, 1986, pp. 42–66, 70–92.

in contesting power and authority, and in defining power relations and the social hierarchy.

The changes in *Wuxing* cosmology epitomize the transformation of political culture in early China. With twentieth-century archaeological discoveries making it possible to reconstruct the history of the Bronze Age, more and more scholars have begun to trace the origin of *Wuxing* to this earliest historical period, to the archaic cosmology of the Four Quarters (*Sifang* 四方). Although a certain symbolic resemblance and some continuity can be found between *Sifang* and *Wuxing*, there are fundamental differences between the two in their structures, functions, and meanings. The early appearance of *Wuxing* can be found in data from the fourth to third centuries B.C., during the transition from the Bronze Age to the imperial era. These early forms of *Wuxing* absorbed certain structural and symbolic features of *Sifang*, and became one of many systems of classification that coexisted in mantic and ritual practices of the time. At this stage, *Wuxing* was loosely defined and unsystematically used. Later, in the course of the formation of centralized empires, it was elevated to become the core of correlative cosmology, the predominant system of classification, and a shared discourse among different interest groups in their political interactions.

Why did cosmology and the political structure go through such fundamental transformations simultaneously? What was the relationship between the cultural and the political aspects of this simultaneous transformation? How did the new meanings of cosmology and the new political institutions construct one another? In asking these questions, this work aims to unveil the process of mutual production of cosmology and empire, of the cultural and the political, in the process of historical change.

Different Approaches and Assumptions

To study the mutual construction of cosmology and empire demands a reevaluation of the methodologies and assumptions inherent in the scholarship on the subject. The close connection between politics – the total complex of power relations in society – and cosmology – the conception of the universe as an ordered system – has long been acknowledged as the fundamental principle of Chinese political order, a principle known in Chinese history as "the union of politics and the doctrine" (*Zhengjiao heyi* 政教合一). But modern disciplinary divisions have split this "union" into a long chain of binary oppositions, those of philosophy versus history, ideas versus institutions, words versus deeds, meaning versus power, culture versus politics, and so on.

Historians' Concern with Origins

Historians of China have studied cosmology primarily as a form of thought, and have been overwhelmingly preoccupied with its origin. According to scholars living during the Han Dynasty, *Wuxing* cosmology was a sacred pattern of the ordered universe, which the sage imitated in creating human culture. These scholars attributed the authority of *Wuxing* cosmology to its sacred and antique origin, as images descended from Heaven and recognized by ancient sages.[11] This idea of the divine and antique origin of *Wuxing* cosmology was taken for granted throughout imperial history, and its influence is still felt today.

One great achievement of modern historical approaches has been the demystification of the origin of *Wuxing* cosmology. Henderson reveals the rise of criticism of correlative cosmology in late imperial times, mostly by Qing scholars of the so-called School of Evidential Research (*kaozheng xue* 考證學).[12] These scholars were engaged in distinguishing authentic classical texts from their interpolations or forgeries, and in establishing their chronology. In so doing, they challenged the passages in these texts about *Wuxing* cosmology with regard to dating and authorship. Their criticism of cosmology, as Bodde has pointed out, was unsystematic, concerned with its textual reference rather than its conception and system.[13] But their sophisticated methods of textual and historical analysis became indispensable tools for the study of Chinese history.

With this heritage, the leading critics of Chinese history and culture in the early twentieth century – such as Liang Qichao 梁啟超 and Gu Jiegang 顧頡剛 – deconstructed *Wuxing* cosmology by revealing the political context within which it was formed and the political motivations of the forgers of *Wuxing* texts or textual fragments.[14] Undermining the myth of its sacred and antique origin, these critics revealed *Wuxing* to be a product of political history. But, limited by the methods inherited from the Qing scholars, their criticism of this cosmology was concerned mainly with its textual reference. Furthermore, they treated cosmology only as a product of textual forgeries carried out by a few Qin and Han compilers motivated by utilitarian and political concerns. Such a conclusion reduced a profound cultural phenomenon – a cosmology that prevailed in all of Chinese society and persisted throughout history – to

11 This theory is recorded in the opening of "Wuxing zhi" in *Hanshu*; see *Hanshu buzhu* 27a (abbreviation *HSBZ*), Wang Xianqian, 1900, rpt. in facsimile, 1983, pp. 1a–b; and *Hanshu* (abbreviation *HS*), Ban Gu, (1962) 1987, p. 1315. For the translation of this statement and detailed discussion of this theory, see Chapter 4 of this book.

12 Henderson, 1984, chs. 7 and 8.

13 Bodde, 1991, p. 102.

14 Liang Qichao, (1926–41) 1986.

a mere lie invented by a few cunning flatterers and usurpers to cover up the real matter of power. Such reductionism implies that politics was the reality, prior to and determinative of cultural life, and that cultural production was a fabricated deception used to mystify that power.

In contrast to historians who give analytical priority to the political, historians of philosophy and science have studied *Wuxing* cosmology primarily as a mode of thinking, a product of mind, or a school of philosophy. While varying in methodology and interests, they have commonly developed a theme suggested by Sima Qian 司馬遷(b. 145 B.C.), attributing the creation of *Wuxing* cosmology to an individual thinker active at the eve of the imperial era, Zou Yan 鄒衍 (305–240 B.C.),[15] and his school of philosophy, which has been retrospectively labeled *Yin-Yang jia* 陰陽家(the school of *Yin* and *Yang*). This school of philosophy is a combination of magic and science according to Feng Youlan, or a school of "naturalists" according to Needham.[16] In arguing against Needham's distinction between Zou Yan's "naturalism" or "protoscience" and Han Confucians' "phenomenalism" or "pseudoscience," Schwartz sees Zou Yan as a pioneer of Han Confucianism, initiating the fusion of cosmology with Confucian values.[17] Xu Fuguan and Li Hansan have each conducted thorough research on classical texts, affirming that it was Zou Yan who transformed the archaic concept of Five Materials (*wu cai* 五材) into *Wuxing*, which then became moving cycles of cosmic energy.[18] These scholars have explored *Wuxing* cosmology in terms of philosophy and science, comparing it to Western philosophy and science.

But treating cosmology as a school of philosophy or a mode of thinking represents another kind of reductionism: it gives ontological or analytical priority to the products of the mind, to ideas, meanings, and thoughts, reducing the social and political enactments of such ideas to mere background. This "mind-centered" approach has further reduced cosmology to pure philosophy, to thought represented in texts, neglecting or obscuring its immense symbolic manifestation in everyday cultural practice and material production. It thus dismembers an immense cultural-political phenomenon, reducing it to the invention of a philosophical school or even a single theorist and limiting the ground for discussion to philosophical texts only. Similar limitations are also found among scholars who are not satisfied with attributing *Wuxing* cosmology

15 The dating of Zou Yan's life is adopted from Qian Mu, (1935) 1985, vol. 32, p. 619.
16 Feng Youlan, 1983, vol. 2, pp. 299–301; Needham and Wang Ling, 1956, pp. 232–53.
17 Schwartz, 1985, pp. 363–9.
18 Xu Fuguan, 1963, pp. 509–87; and Li Hansan, 1981, pp. 30–5, 51–62.

to Zou Yan as the sole innovator. Henderson has tried to trace multiple sources for Chinese correlative cosmology, yet all the sources he traces are confined to the realm of philosophical schools and the syncretism of Han philosophers.[19]

Both historical approaches to Chinese cosmology, that of political history and that of the history of philosophy, have made tremendous contributions to the building of a solid foundation of textual analysis, which has made the present study possible. However, neither approach has proven sufficient by itself to explain how and why cosmology became a pervasive and profound cultural phenomenon, or to show the interrelation of the simultaneous transformations in cosmology and political structure in early China.

Classical Anthropology and Sinology

Structural anthropologists have led the study of Chinese cosmology in an opposite direction. Instead of treating it as a conscious invention of philosophers at a certain time in history, they see correlative cosmology as a mode of thinking universal to primitive cultures or even to all cultures, one that is particularly enduring in Chinese civilization. Instead of focusing on the problem of origins, anthropologists have been concerned with the structure and symbolism of cosmology and its connection to society and culture as a whole.

However, in their pursuit of a holistic reconstruction of the sociopolitical and the cultural, structural anthropologists repeat the chain of dichotomies between ideas and institutions or between the cultural and the sociopolitical, giving priority to one side or the other. Attributing ontological priority and causality to the social realm, Durkheim sees correlative classifications in general as "reflections" or "imitations" of preexisting social structures. He sees Chinese cosmology as one such reflection, even though a clear link between the social system and classification is unsupported by the evidence and therefore still undemonstrated.[20] In contrast to Durkheim's social structuralism, Lévi-Strauss and Eliade attribute ontological priority to the symbolic structure of the human mind. They conceptualize cosmology as a given structure of mind common to most archaic peoples, which social reality imitates and repeats.[21] Wheatley further applies this theory to the study of Chinese

19 Henderson, 1984, pp. 28–46. 20 Durkheim and Mauss, 1963, pp. 73–4.

21 Lévi-Strauss, (1962) 1966; Eliade, 1949, ch. 1; as summarized and developed in Wheatley, 1971, pp. 416–18.

cosmology, contending that the reality of Chinese cities was a function of the "imitation" of a preexisting "Celestial Archetype" or "Symbolism of the Center."[22]

These two anthropological approaches are analytically polarized, viewing cosmology as either the reflection of a prior social structure or a preexisting conceptual model that social reality imitates. Both approaches reevaluate the connection between cosmology and social reality, and both conceive them holistically; but the unity each constructs preserves the dichotomy between the cultural and the sociopolitical, serving as a static and eternal structure that transcends history. Consequently, the question of cosmology's historical development becomes irrelevant.

Some sinologists have adopted the principles of structural anthropology, especially the approach represented by Lévi-Strauss and Eliade, but have rejected its conclusion of a universal primitive mode of thinking. Rather than ascribing correlative cosmology to primitive cultures, for example, Granet stresses that Chinese cosmology is a highly ordered system based on the logic of numbers, which functions as classification and as protocol. Chinese cosmology, according to him, is a logical unfolding of structural principles of symmetry and centrality.[23]

Granet's insight has stimulated many sinologists to further develop his thesis and approach in their own research. For example, Needham carries out the theme that Chinese correlative thinking was not primitive thinking in the sense that it depicted not an illogical or prelogical chaos, but rather a picture of a highly and precisely ordered universe.[24] Bodde employs Granet's structural principle of symmetry and centrality in explaining the growth of *Wuxing* cosmology.[25] And Major specifically traces the structural origin of *Wuxing* cosmology to the numerology of the magic square.[26]

The most influential sinologist in this direction is A. C. Graham, who combines a sophisticated structural analysis with a notion of historical development. Graham challenges the classical structuralism that defines correlative thinking as a stage of prelogic at a lower level of evolution of human intelligence, one belonging especially to China or to primitive cultures, and argues instead that in both China and the West we find different levels of thinking, correlative and logical, in philosophy and protoscience.[27] Through structural analysis, he demonstrates that like correlative thinking in general, which is rooted in the interplay between linguistic "paradigms" and "syntagms," the Chinese cosmology of

22 Wheatley, 1971, p. 418. 23 Granet, (1934) 1950.
24 Needham and Wang Ling, 1956, p. 286. 25 Bodde, 1991, pp. 103–21.
26 Major, 1984. 27 Graham, 1986, pp. 3–15; 1989, pp. 315–19.

Yin-Yang Wuxing evolved from a universal structural scheme of binary oppositions.[28] Yet unlike classical structuralists, who stop at the universality of correlative thinking, Graham includes a historical notion in his structural analysis by tracing the early development of *Wuxing* cosmology. He points out two important historical phenomena. First, during the classical period (before 250 B.C.), correlative cosmology prevailed only in protoscience, and philosophers were indifferent if not hostile to cosmologies. Second, *Wuxing* cosmology was initiated by Zou Yan when he integrated the concept of Five Processes from protoscience into his political theory and was then systematized and elevated by Han philosophers to become the prime cosmology, again for political purposes.[29]

While adopting the vocabulary and concepts of structural anthropology, these sinologists retrieve the basic themes and assumptions of historical approaches described earlier. Like Chinese historians Gu Jiegang and Feng Youlan, Western sinologists ponder the question of origin, and many agree that *Wuxing* cosmology was invented by Zou Yan and completed by Han philosophers. For evidence, they have commonly confined themselves to texts, mostly classical and philosophical texts, while with regard to subject matter they see cosmology as a form of human intellect.

Culture, Ideology, and Power

Beyond Disciplinary Boundaries

The disciplinary boundaries discussed earlier have divided the single historical process of forming and transforming cosmology and the political system into two separate subjects. Various disciplines have addressed the relationship between the two subjects only at the level of causality or ontological priority – that is, which one is prior to and thus determinative of the other. Institutional history gives such priority to forms of government and bureaucracy, viewing cosmology as the invention of certain individuals with political motivation to justify the existing system. The history of philosophy, on the other hand, studies cosmology as a pure mode of thought, one that is permanent and universal, and sees political use of this cosmology as corruption of the structure. Structural anthropology is also concerned with the causal relationship between cultural ideas and sociopolitical institutions, attributing causality and onto-

28 Graham, 1989, pp. 319–25, 331–56; 1986, pp. 16–66.
29 Graham, 1989, pp. 325–30; 1986, pp. 70–92.

logical priority to one or the other. It has sacrificed history in the reconstruction of the relation between the cultural and the political, ignoring human agents, practices, and the process of change.

The present work breaks down these disciplinary boundaries and reconstructs the dialectic mutual construction of cosmology and the empire, that is, of the cultural and the political realms. In order to articulate the need for a total analysis of this phenomenon, the concept of a cultural-political "totality," borrowed from the anthropologist Stanley Tambiah,[30] helps us to understand *Wuxing* cosmology as a unity of structure and events, of conceptions and actions, and of continuity and transformation. This unity, nevertheless, does not suggest a total history, nor a homogeneous equilibrium in a Durkheimian sense, nor an integrated system of ideology and class domination in a Marxist sense, nor a hegemony or homogeneity. On the contrary, it depicts a single, but highly dynamic, process of cultural-political change that is constituted by fragmentary events, by conflicts and disputes, and by contests for power and control.

To study cosmology as such a cultural-political "totality" requires an interdisciplinary approach that focuses its analysis upon the connection of the disciplinary boundaries, studying the process of interrelation and mutual construction of the cultural and the political rather than treating them as separate entities. The interdisciplinary approach also incorporates perspectives, methodologies, and materials from history, anthropology, archaeology, and philology, making connections between archaeologically discovered material culture and written records, between popular practices and state ideology, between philosophical debates and historical events, and between the symbolic construction of cosmology and the institutional construction of empire. Such an analysis is both a historical anthropology and a cultural history of early China. By applying such methods to the study of early China, I also mean to bring the history of early China – a field that is still seen as a unique and exotic "other" in the West and that remains isolated, accessible only to a limited number of highly specialized scholars – into general theoretical discussions about culture and power.

The Analysis of Culture, Ideology, and Power

Investigating ancient Chinese cosmology as an intrinsic component of power and as a discursive production of empire contributes to the the-

30 Tambiah, 1985, pp. 1–7.

oretical discussion of ideology and power. Classical Marxist analysis has established that ideology is a system of beliefs or ideas that functions to sustain relations of domination. The most influential writing of Marx and Engels on ideology is found in *The German Ideology*, in which ideology is defined as the intellectual production of the dominant class: "The class which has the means of material production at its disposal, has control at the same time over the means of intellectual production . . . The ruling ideas are nothing more than the ideal expression of the dominant material relations, the dominant relations grasped as ideas; hence ideas of its dominance."[31] The relations of production constitute the "real foundation," to which correspond "definite forms of social consciousness."[32] Thus the dominant ideology serves to legitimate the existing system of class domination by way of producing social consciousness – a deception fabricated by the dominant class to cover up actual economic exploitation.

Sociologist Max Weber rejects the economic interpretations of ideology in Marxism. Unlike Marx, Weber sees values or beliefs as anything but secondary to the economic or political. As Weber illustrates, the Protestant ethic provided indispensable dynamism for the development of capitalism in Europe.[33] Like Marx, however, Weber is also concerned with the relationship between ideas and domination. He contends that different systems of domination attempt to establish belief systems to legitimize themselves, and proposes three basic types of legitimate domination, those based on rational, traditional, and charismatic grounds.[34]

Not content with classical Marxist economic determinism and Weberian ideal types, neo-Marxist theorists have developed new dimensions for the analysis of ideology, each a new way of examining relations between ideas and domination. They extend the concept of ideology far beyond the classical boundaries of "beliefs" and "consciousness" and the immediate economic interests of the dominant class. For example, Gramsci's concept of "hegemony" extends the notion of ideology from immediate economic interests to the political leadership and cultural domination of the ruling class. Controlling material production alone cannot establish class domination; the ruling group must also take leadership in the production of culture. Gramsci thus enriches and reinforces the Marxist idea of ideology by adding symbolic production and

31 Marx and Engels, (1845–6) 1965, p. 61.
32 Marx and Engels, 1958, vol. 1, p. 363.
33 Weber, (1958) 1976.
34 Weber, 1978, vol. 1, chs. 1 and 3.

cultural domination to the classical Marxist view of the ruling class's domination of material production.[35]

Gramsci's concept of hegemony was further developed and modified by anthropologists. Some current anthropologists, with their roots in fieldwork and a tradition of focusing on non-European and nonelite cultures and peoples, have integrated the anthropological concept of "culture" and the Marxist concept of "ideology," incorporating culture, symbolic systems, and language into the analysis of the relations of domination. The Comaroffs criticize both traditional anthropological conceptions of culture, which are "neutral and above history," and the Marxist concept of ideology, which neglects "the meaningful bases of consciousness and the expressive forms of ideology" – in other words, culture.[36] They raise the question of how culture constructs and is constructed by ideology and thus is involved in power relations. By introducing the concept of "habitus," Bourdieu reveals how the symbolic system of classification that works below the level of explicitly formulated ideas functions to reproduce the structure of social order and class domination.[37] Hebdige, like Bourdieu, sees the politics of ideology functioning as a symbolic system of signs that affect people "as structures" rather than "via their consciousness." But unlike Bourdieu, whose "symbolic classification" represents a single, unitary, totalized structure, Hebdige defines "culture" as a field of class struggles taking the form of battles between the dominant hegemonic culture and resistant subcultures.[38] Scott and van Onselen reject Gramsci's notion of "hegemony" altogether, revealing the existence of unorganized and unarticulated resistance among subordinate classes in the form of an implicit language of symbolic activity in everyday life.[39]

While current anthropology has greatly enriched the analysis of ideology with the notions of culture and symbolic meaning, it is still by and large handicapped by an inherent reductionism. As already shown, there are two basic models for the analysis of ideology. One is what Donham calls the "power/ideology" reproductive model, which sees ideology as the reproduction of productive inequalities.[40] This model has been articulated in Gramsci's "hegemony" and Bourdieu's "habitus." The other model can be identified as the dominance/resistance dichotomy,

35 The concept of "hegemony" came originally from Antonio Gramsci's famous prison notebook (1971). The concept has been used in various fashions by many neo-Marxist theorists, such as Habermas and Marcuse. For a critique of the limitation of "hegemony," see Scott, 1985, 314–50.

36 John L. Comaroff and Jean Comaroff, 1987; 1991; Jean Comaroff, 1985.

37 Bourdieu, 1984, pp. 170, 465–71.

38 Hebdige, 1979.

39 Scott, 1985, pp. 304–50; Onselen, 1976, pp. 227–44.

40 Donham, 1990, pp. 204–5.

which treats ideology as a battlefield between dominance by the ruling class and resistance by the ruled. This second model is reflected in Hebdige's "subculture," Scott's "weapons of the weak," and the Comaroffs' "symbolic struggle."

Both models, however, suffer from reducing the complexity of human experience in the domain of meaning and power to either a unitary and static system of dominance or a dichotomy of dominance/resistance. As Sally F. Moore forcefully asserts, the "power/ideology" reproduction model "gives those phenomena the appearance of operational inseparability and congruent effect. But in ethnographic fact they may (or may not) be mutually reinforcing and reiterative." Therefore, "such abstractions are not suited to addressing diversity, uncertainty, and transformation." The "dominance/resistance" model is equally simplifying and reductionist. It "gives analytic life to a mythic reduction of the complexity and multifacetedness of human thought. There can be many more than two sides, many more than two postures, many more than two ideas of 'reality'."[41]

The most enlightening perception of the anthropological analysis of ideology lies in treating culture as intrinsic and constitutive of power. It is this perception that is employed here in investigating how Chinese cosmology, the commonly recognized foundation of Chinese culture, is intertwined with and constitutive of the power relations of the empire. This study rejects the reductionist models of reproduction of "ideology/power" and "dominance/resistance." Rather than describing a unitary Chinese culture as a "hegemony," a "habitus," the legitimization of rule, or a dichotomy of dominance/resistance, I endeavor to describe the multifaceted nature of ancient Chinese cosmology, the uncertainty and fluidity of its meaning, the diversity in power relations, and the transformation of political culture.

This study undertakes a "genealogy" of cosmology in part because it is a significant cultural resource for the politics of identity and nationalism today. Adopting Foucault's concept of genealogy,[42] my examination of early Chinese cosmology is aimed at tracing not its origin, its progress, or its deep meaning as the structure of Chinese mind or civilization, but rather its complex and diverse applications in practice, the social contest and conflict that surfaced through contradictions in the interpretation of cosmology, and finally, the power relations of the early empires of China that constructed and were constructed by the cosmological discourse. By treating cosmology as a "discourse," I mean to reject

41 Moore, ed., 1993, pp. 3, 9.

42 For the analysis of Foucault's approach to "genealogy," see Dreyfus and Rabinow, (1982) 1983, pp. 104–17.

the essentialistic view of cosmology, culture, and Chineseness. Instead of treating it as a system of abstract ideas or signs, as the concept of ideology suggests, I treat cosmology as a "discourse" embodying both theory and practice, and encompassing both domination and contention. Using Foucault's term, such discourse consists of "historically analyzable practices" that systematically produce subjectivity and power relations.[43]

Shifting the analysis of culture and ideology from unitary systems of domination to a multifaceted and complex phenomenon of contention, this study also reexamines the concept of power. It does not see power as a metaphysical or ontological substance that can be processed and owned, that the ruler monopolizes to dominate the ruled. Instead, it examines power in early China as a set of complex and dynamic relations, as what Foucault calls "the extremely complex configuration of realities," as what "designates relationships between partners," and as "a mode of action upon the action of others."[44] The complex and dynamic nature of power relations in early China has been concealed by the claims of imperial sovereignty, the institutional definition of emperorship, and the cosmological claims of a single source of authority and power. To demystify the ontological claims of power and decode the concealed dynamism and complexity of power in the case of early China is a major goal of the book. This will be done by revealing the political contest over imperial sovereignty, emperorship, and imperial order that is veiled by a commonly shared cosmology.

Theory, History, and Modernity

Historical Anthropology and Cultural History

Analyzing the dynamic mutual transformation of culture and power, this study reflects the recent shift in both anthropology and history toward a time-oriented historical anthropology and a theory-engaging cultural history. And this shift has resulted from the changing conceptions of culture, ideology, and power just discussed.

The emergence of historical anthropology is a result of the rethinking of culture, the cardinal concept of anthropology. Because culture is no longer seen as exotic customs and festivals but as inherent to power and to history, the study of culture must include human agents and actions, dynamic changes in power relations, and the (re)invention of power's cultural forms. Beginning with the structuralism of Durkheim and Lévi-Strauss, through the symbolic anthropology of the 1960s and

43 Ibid., pp. 62, 250. 44 Ibid., pp. 217, 221.

the structural Marxism of the 1970s, the antithesis between history and structure has been repeatedly reinforced, and the analytical priority given to structure over history has been held by many to be the disciplinary identity of anthropology. During the last two decades, however, leading anthropologists have started to reveal the analytical shortcomings of such an ahistorical treatment of systems or structure. The "practice approach," developed by Sahlins and Bourdieu, represents one attempt to implicitly unify history and structure. This approach introduces to the analysis of structure – whether economic, political, or symbolic – the concept of human agents, events, and actions, all constrained by, yet at the same time reproducing and transforming, the structure.[45] As Sahlins says: "'Structure' – the symbolic relations of cultural order – is an historical object."[46]

While the practice approach of Sahlins and Bourdieu has provided a promising model for historical anthropology, its limitations have been recognized. One such limitation is that "structure" in Sahlins's and Bourdieu's works is still treated as a given, preexisting condition; the process of creating meaning and constructing a system has been largely overlooked. Furthermore, the source of change in these works is mostly described as coming from individual reaction to external events rather than from internal dynamics, as exemplified by Sahlins's analysis of Hawaii's history.[47] To overcome such limitations, anthropologists have undertaken long-term histories in an attempt to uncover the internal dynamics and systematic changes of structure. For example, Tambiah, tracing the contemporary political system in Thailand to early Buddhism, uncovers "a recurrence of structures and their transformations in systematic terms," and "dynamics of polities" behind the cosmology and doctrine.[48]

Another limitation, a consequence of the first one, is that the practice approach is an individual-centered model of change. As Moore points out, such a model understands cultures and societies "to be continuously produced and transformed through the medium of the generic individual," and thus is "of little assistance in addressing nonuniversal regularities." Moore further suggests that instead of choosing the extremes of scale – the world system or the generic individual – social analysis of dynamic change be directed to the intermediate levels, the dimensions of "relations," "resources," and "representations."[49]

While anthropology has moved to embrace history, history itself has comprised culture. Because culture is inherent to power, "power" can

45 For representative works of this approach, see Bourdieu, 1977; and Sahlins, 1981.
46 Sahlins, 1985, p. vii. 47 Ibid.
48 Tambiah, 1976, pp. 5, 123. 49 Moore, 1986, pp. 9, 328.

no longer be defined as the possession of institutions or the state, but extends into all domains of culture in everyday social life, such as language, signs, the family, and ritual. In one word, history is culturally constituted. The rise of cultural history since the late seventies represents this new understanding of history. Cultural historians turn away from rational institutions of law and bureaucracy to study signs, language, discourse, visual image, ritual, and artifacts. To analyze such subjects, cultural historians also turn to theories of culture. Gramsci's "hegemony," which emphasizes the cultural domination of the ruling class, inspired not only anthropological concepts, such as Bourdieu's "habitus" and Scott's "weapons of the weak," but also masterpieces in cultural history such as E. P. Thompson's history of the English working class.[50] Some cultural historians directly adopt anthropological theories, such as the practice approach of Sahlins and Bourdieu or Geertz's concept of culture and his method of "thick description." Geertz, in accord with the practice theory, sees human behavior as "symbolic action" and culture a semiotic text in which the symbolic meanings of the action can be read. Since culture is embedded in details of daily life, the task of ethnographers is to interpret, or thickly describe, culture through microscopic reading of the symbolic meaning.[51] Many cultural historians embrace Geertzian interpretive methods in doing their "ethnography of the past" or "history in the ethnographic grain." Others, however, warn that Geertz "never confronted the issue of power"[52] and that Geertzian cultural history has jeopardized causal explanation altogether by replacing it with interpretation that decodes meaning.[53]

This study of early China addresses the fundamental question of both anthropology and history concerning how to analyze culture and power in relation to one another and in the process of change. It seeks to transcend the limitations of the practice approach that sees cultures and societies as given structures. Investigating the mutual construction of Chinese cosmology and empire in their formative stages, this book examines both the political dynamics – power contests and political changes – that transformed cosmology and the cosmology that was actively created and mobilized as a discursive expression of power

50 Thompson, 1968. 51 Geertz, 1973, pp. 10, 14, 20–21.

52 Dirks, Eley, and Ortner, eds., 1994, p. 22.

53 Appleby, Hunt, and Jacob, 1994, pp. 219–23. Geertz himself best distinguished the difference between his cultural analysis and causal explanation: Cultural analysis "is not an experimental science in search of law but an interpretive one in search of meaning." "Cultural analysis is (or should be) guessing at meanings, assessing the guesses, and drawing explanatory conclusions from the better guesses." (1973, pp. 5, 20).

relations. Furthermore, instead of focusing on either the "generic individual" or the imperial state as the totalizing system, this study locates the mechanism of change in conflicting sociopolitical groups, in their asymmetrical relations and their competition in mobilizing symbolic and cultural resources. Finally, distinguishing different uses of the same cosmological discourse by these sociopolitical groups, this study attempts to return the Geertzian interpretation of culture to the field of politics, power, and social change. This work, therefore, explores both the cultural and discursive constructions of power and subject, and the social agent's active transformation of the cultural discourse in political change.

Theory and Comparative Studies

The advance of social theory is achieved by comparing numerous unrepeatable histories and localized ethnographies, and by particular textual analysis. Comparative study, therefore, is indispensable for theoretical formulation. Peter Burke points out that not only theoretical generalization but also the particularization of history "depend on comparison," as evidenced in the great achievements of comparative history.[54] The history of China, with its continuity and rich documentation, has great potential to contribute to, challenge, and advance existing social theory, and to enrich the comparative studies of history. Kwang-chih Chang, the leading archaeologist of China, has been instrumental in forging such a theory regarding the historical anthropology or cultural history of China. He treats Chinese culture, whether art and writing or civilization as a whole, as intrinsically constructive to political power. His works integrate social theory into the study of ancient China and address general questions of human history, using data from ancient China to test, challenge, and modify existing social theory. He overturns the Western-centered stereotypes of ancient China as the unique exception to the evolutionary scheme – the "Oriental" or "Asiatic" society in classical Marxist literature, the "patrimonial state" in Weber's typology, a "hydraulic society" in the theory of Wittfogel.[55] Furthermore, by comparing ancient China to ancient Mesoamerica and other civilizations, he concludes that "China is far from being unique – rather, its pattern is repeated within many other ancient civilizations."[56]

54 Burke, 1992, pp. 22–28. 55 Kwang-chih Chang, 1983, pp. 125–9.
56 Kwang-chih Chang, 1986, p. 419.

Chang, using data from China's Bronze Age, has established a model of actively developing rather than passively applying social theories. While Chang's works study Bronze Age China, imperial China is still greatly in need of a similar approach using historical anthropology. With its two thousand years of well-chronicled dynastic histories and massive records beyond historiography, China's imperial history is still by and large isolated within Western sinology and Chinese *Guoxue* 國學 (the study of national heritage). This work is a preliminary endeavor to break such isolation, to bring the data of early imperial China into the comparative study of empires and royal ideologies, and to use this data to enrich theoretical discussions.

This study of Chinese cosmology and early empires contributes to the comparative studies of imperial formation. Since Kantorowicz's *The King's Two Bodies* laid the foundation,[57] the comparative study of kingship, imperial formation, and royal ideology has developed an impressive body of scholarship. Within this scholarship, two works bear particular comparison to the present study of early China. Anthropologist Ronald Inden has developed the concept of "imperial formation" based on his study of India, but he means thereby to provide a general conceptual tool for the study of other regions of the world. With the concept of "imperial formation," he depicts "a complex polity consisting of overlapping and contending agents related to one another in a 'world' whose spokesmen claim universality for it."[58] Inden's depiction is very helpful for demystifying the Chinese claim, made over two thousand years of imperial history, that the unified empire of China is an entity, an essence, or a substance. I have found in my research on early imperial China a phenomenon similar to that of the "imperial formation" of India, a claimed unified empire that was nothing more than a body of relations tying together contending political forces and interest groups.

Tambiah, who has greatly contributed to the comparative study of kingship, argues that the Buddhist concept of universal king, as a form of ideology, was not the ruler's mystification used as an instrument of exploitation and domination, but rather "a model for and model of political conduct."[59] This approach to the relation between ideology and practice helps us to escape the "legitimation" or "reproduction" model of analysis and to decode Chinese emperorship – which has been seen as a means of legitimizing the emperor's power – as a discourse through which various competing political forces contested with one another in defining social norms.

57 Kantorowicz, 1957. 58 Inden, 1990, p. 2. 59 Tambiah, 1987, p. 35.

Ancient China and Modernity

Situating this research in a larger theoretical, methodological, and comparative context beyond the specialized field of sinology, we see that the case of early China has a lot to contribute to the development of social theory. But this is important not simply for the sake of abstraction, for developing theory per se. Rather, reexamining China's past in relation to social theory is a particularly urgent task for today's Chinese national self-awareness. The isolation of China has resulted from both the "closed door" policies of the Chinese communist state during the cold war era, and the Eurocentric construction of the oriental "other." Being accustomed to seeing itself, and being seen by others, as a unique civilization is one of the obstacles to modernity that China faces today.

The "Chinese cultural heritage" that has been mystified throughout history continues to be conceived of as a metaphysical and ontological substance. This heavily mythologized, essentialist view of Chinese "tradition" is the basis for various conflicting positions in the political debate on China's transformation in the modern world. Some contemporary Chinese thinkers see Chinese civilization – its essence symbolized by "dragon," the Yellow River, or a "yellow" civilization – as a "deep structure" or a "national psyche," causing China to be unprogressive, static, or stuck in its own cyclical repetitions.[60] They see the only hope for China to escape the fate of being permanently locked into this "Oriental despotism" as Westernization, which can be achieved in China only by "one very special group: its intellectuals," who "can communicate directly with the civilization of the sea" – what they call the "blue" civilization of the West.[61]

Some other contemporary Chinese, by contrast, affirm that "what's truly tragic" for China today "is not the weight of tradition but the absence of it," China's "traditionless void."[62] A similar attempt to revive Chinese tradition as a path to modernization is the so-called "new authoritarianism" or "neoconservatism" propagated by some intellectuals, calling for an "enlightened dictator" who could introduce democ-

60 This voice is forcefully articulated by a group of academics, writers, and television directors in the six-part television documentary *Heshang* (River elegy). References are to Su Xiaokang and Wang Luxiang, 1988. For analysis and partial translation, see Barmé and Jaivin, eds., 1992, pp. 138–64.

61 *Heshang*, episode six, "The Color Blue," translation adopted from Barmé and Jaivin, eds., 1992, pp. 155–6.

62 This view is voiced by He Xin, a self-proclaimed "cultural conservative" writer, whose statement is translated in Barmé and Jaivin, eds., 1992, p. 162.

racy into China without paying the price of anarchy.[63] This search for an "enlightened dictator" shockingly resembles the two-thousand-year-old practice of longing for an "enlightened emperor"; both attribute the responsibility for social well-being to a single exemplary human agent. Whether it is the death of the "yellow" civilization or the revival of authoritarianism, both opposing positions add yet another coating of mystification to the concept of "Chinese culture," reinforcing its imagined metaphysical and ontological essence.

Demystifying such an essence, this work reveals that what was commonly held to be the "essence" of Chinese culture and civilization – the empire and emperorship – was a continually constructed and contested discourse rather than the substance of power. The power relations and contestation disguised by the "totalitarian oriental despot" were not, in fact, fundamentally unique or exotic when compared to histories of other human societies. It is a double defect to use the past ideological claims of a unitary and essential Chinese culture to analyze modern China's reality. And it is doubly deceptive to reiterate the mythology of a single dictator – whether past emperors or a modern despot – as constituting the only active and responsible human agent, a myth that serves only to veil the exceedingly complex power contest going on today.

Outline of the Book

Following this introductory chapter, Chapters 2 and 3 analyze the mutual construction of cosmology and political power during the historical transition from the Bronze Age to the imperial era. This analysis draws evidence from different fields, including material cultural remains, oracle bone and bronze inscriptions, and texts recently unearthed as well as transmitted. Chapter 2 demonstrates how, during the Bronze Age, the ruling clans of Shang and Western Zhou (ca. 1700–771 B.C.) conceived their authority and asserted their political dominance through a cosmology of Four Quarters (*Sifang*). This cosmology conceptualized the universe in terms of four quarters surrounding a center, the center being conceived of as the king's body and his ancestral line, through which the world of the gods and the world of human beings communicated. By occupying such a sacred center and monopolizing the axis of connection to the divine through ancestor

63 Even radical activists such as Dai Qing and Liu Xiaobo have to varing degrees supported the new authoritarianism; for translations and discussion on the new authoritarianism, see Barmé and Jaivin, eds., 1992, pp. 184–90; Dittmer and Kim, eds., 1993, p. 148.

worship, the king achieved both political domination and divine authority. This sacred and static centrality of the king and his ruling clan was defined by the alienness, otherness, and inferiority of the *fang* polities in political geography, and by the four quarters of the cosmos.

Chapter 3 shows how the transformation of cosmology from the Four Quarters (*Sifang*) to the Five Phases (*Wuxing*) was intrinsic to the political transition between the Bronze Age and the imperial era, that most drastic historical change that occurred during the Warring States period (ca. 500–221 B.C.). Various rising political groups of this time – such as religious specialists, military professionals, scholars, and bureaucrats – all served a new kind of territorial power and mobilized a discourse of *Wuxing* (Five Phases) cosmology in constructing new power relations. Dismantling the old power construct, they negated the cosmological expression of the political and religious centrality of the ruling clan, replacing the notion of a sacred and eternal center defined by the Four Quarters with a new system based on dynamic interactions of five cosmic phases. They used this *Wuxing* cosmology to build direct correlations between the human world and the cosmos, superseding the medium of the royal ancestors in communication between Heaven and Man. Direct correlations between Heaven and Man not only created new sources of divine authority, but also constructed a concept of human sovereignty for the unified empire that was forming.

Chapter 4 studies *Wuxing* cosmology as a political discourse during the first empires of Qin and Han in the last two centuries B.C. It analyzes how *Wuxing* discourse embodied tensions among competing social forces over the transmission of power, the sociopolitical order, and imperial sovereignty; and how such tension in turn transformed *Wuxing* cosmology in dynamic discursive practices. In introducing the most comprehensive representation of the cosmological discourse of the Han Empire – "Wuxing zhi" 五行志 in *Hanshu* 漢書 – the chapter singles out two structural principles of this text – cycles of cosmic movement and the cosmological structure of Earth, Man, and Heaven. Analyzing the contradictions in the first principle, this chapter relates the cosmological debate over imperial sovereignty that was expressed in two competing cosmological orders – the conquest cycle of *Wuxing*, representing a sovereignty based on force and punishment, and the generation cycle representing a sovereignty based on ethical principles, rituals, and hierarchies. Investigating the second structural principle – the cosmological relationships among Earth, Man, and Heaven – Chapter 4 further unveils a long process in which a cosmology of conquering force was transformed into a moralized cosmology. These opposing cosmological constructions of imperial sovereignty were put into practice by two exemplary emperors. China's First Emperor, the first emperor of Qin, realized sovereignty based on the cosmology of force and punishment. By contrast, Wang Mang, the greatest practitioner of Han Confucian ideals, enacted an emperorship defined by the moralized cosmology.

Chapter 5, the final chapter, focuses on the practice of omen interpretation that was carried out within *Wuxing* discourse, and how such discursive practice formed and transformed emperorship, the pivot of imperial power relations. It demonstrates that the commonly shared discourse of emperorship, which designates the emperor the single responsible agent for order and disorder in the universe and human society, was produced and reproduced through vital political contest. One such contest was carried out between scholar-officials, who defined and confined emperorship by systematizing cosmology and monopolizing moral authority, and the emperor and his hired religious specialists, who resisted the constraints of moral authority by seeking direct contact with the divine world outside of the systematized cosmology. Another contest over emperorship was that between royal nobles, whose power was based on local kingdoms and blood ties to the emperor, and the scholar-officials running the centralized government. In this case, the two sides competed by attempting to influence the emperor, in a heated cosmological debate over relations between Heaven and Man and modes of rulership based on different interpretations of cosmology. These theoretical arguments were directly employed in the political struggle over two opposing forms of imperial order – centralization and pluralism. The conclusion of the chapter points out the fundamental changes that occurred in the rulership of China between the Shang model of the king as the center of the universe and the Han model of the emperor as the pivot of the cosmos and empire.

2

Sifang and the Center: The Cosmology of the Ruling Clan

Introduction

Applying a structural analysis to ancient Chinese thought, both Angus Graham and Derk Bodde see the emergence of the *Wuxing* (Five Phases) system as the result of inserting centrality into older symmetry oppositions, which in turn caused a shift from four components to five, a phenomenon first seen in the texts of the fourth to second centuries B.C.[1] But even during the Shang period (ca. 1700–1045 B.C.), long before the age represented by such texts, the symbolism of centrality in symmetry oppositions already prevailed in oracle bone and bronze inscriptions and in material cultural remains. Although no evidence of the concept of *Wuxing* during the Shang period has been found, the Shang cosmology, structured on *Sifang* 四方 (Four Quarters) and a center, fully manifests centrality through symmetry oppositions. It is based on this structural similarity that scholars of Shang civilization have argued that *Sifang* (Four Quarters) cosmology was the antecedent of *Wuxing* (Five Phases). As early as 1915, Luo Zhenyu suggested that the Shang period had the concept of the five *Di* 帝.[2] Hu Houxuan in 1944 suggested that the five *fang* (or quarters), composed of the four *fang* and a center that were found in Shang oracle inscriptions, was the origin of *Wuxing* cosmology, supporting this point with a linkage between Shang inscriptions and later texts.[3] Sarah Allan's recent work has elaborated upon Hu's thesis, concluding that *Sifang* and a center formed a *ya* 亞-shaped notion of the earth, from which *Wuxing* theory originated.[4] Li Xueqin and Chang Zhengguang found the origin of the basic correlation of *Wuxing* 五行 (Five Phases) cosmology in Shang's correlation of the four

1 Bodde, 1991, pp. 102–21; Graham, 1986, pp. 49–51; 1989, pp. 342–4.
2 Luo Zhenyu, 1915, as quoted in Hu Houxuan, 1944c, p. 1a.
3 Hu Houxuan, 1944b, p. 1a; 1956.
4 Allan, 1991, pp. 76, 100–2.

seasons and seasonal activities with *Sifang*.[5] Wang Tao and Shen Jianhua further traced the origin of *Wuxing* to the color symbolism in Shang – the use of various colors of jade and sacrificial animals in rituals.[6]

This prolonged search for the origin of *Wuxing* in Shang has established two points – that a certain structural continuity between Shang cosmology and the *Wuxing* system did exist, and that the spatial notion of *Sifang* is of primary importance as one of many possible origins of *Wuxing*, including the correlation of seasons and directions, the numerology of five, and color systems.

While the structural continuity between *Sifang* cosmology and the *Wuxing* system has been well recognized, this study attempts a systematic investigation of the fundamental changes in the meanings and functions of these similar structural principles in different social and historical contexts. David Keightley has warned of the danger of overlooking the holistic historical context of early Chinese cultural phenomena in seeking origins for subsequent ideas. He asserts that "any study of early China involves not just the origins of what we know as later Chinese civilization, but also the origin of early civilization in China," since "not all religious manifestations from this early period necessarily left their mark on later Chinese culture."[7] Complementing Keightley's view, I argue that even in instances where cultural phenomena did continue in later Chinese culture, as in the case of *Sifang*, their meanings and functions could be totally transformed or reinvented by social agents in historical process. There is, therefore, another danger in seeking origins for cultural phenomena, that of neglecting the transformative process involved in borrowing an earlier cultural code for later purposes.

It is the *transformation* of cosmology from *Sifang* to *Wuxing* – the changes in meaning and function hidden beneath their structural continuity – that occupies this study, rather than the origin of *Wuxing* in *Sifang* per se. As a first step in investigating this transformation, this chapter discusses *Sifang* cosmology in the social and political context

5 Li Xueqin, 1985; Chang Zhengguang, 1989.

6 While Wang Tao and Shen Jianhua found some evidence matching Shang color symbolism with that of *Wuxing* (see Wang Tao, 1993a; 1993b; Shen Jianhua, 1993), we must also consider the evidence showing the differences between the two color systems in ritual context. The colors found in Shang inscriptions, which are not systemized into five, were not correlated to the five *fang*. For example, yellow oxen were used in *liao*-sacrificial rites, offered equally to the East, the West, and the South (*Heji*, 14313a, 14314, 14315). The black animals, with "black" and "drought" sharing the same character, were most often used in rain-seeking rites in which *Sifang* was not even mentioned. White animals were mostly used in *liao*-sacrifices to the ancestors, and also in rain-seeking rites, but not in the sacrifices to *Sifang*. We can say only that the meaning of colors in the Shang is still not clear, but that they clearly did not match the directions in the *Wuxing* cosmology of Han times.

7 Keightley, 1995.

of China's Bronze Age – the Shang and Western Zhou. It examines the *Sifang* cosmology at multiple levels – as a spatial and geographic concept, as a cosmological structure for classifying all forces in the universe, and as a ritual structure for communication with the spiritual world. It first situates the cosmology in the political context of Shang, establishing the essential role *Sifang* cosmology played in defining the centrality of the politically dominant Shang clan, thereby functioning as a constructive force shaping power relations. The chapter then reveals how *Sifang* cosmology was used to organize ritual and political actions, ordering the everyday reality of the Shang world in terms of time and space. Finally, the chapter discusses the continuity and transformation of *Sifang* cosmology across the divide between Shang and Zhou, when another ruling clan took over the Shang hegemony.

The evidence presented here comes primarily from the Shang and Western Zhou periods, including oracle bone and bronze inscriptions, archaeological data, and remains of material culture such as bronze and jade objects. For the discussion of Shang cosmology, I rely solely on the Shang data, and on the coherence among these different types of data. Since transmitted texts containing information about Shang cosmology are mostly dated to the fourth century B.C. and later, I choose not to use these texts to interpret the meaning of Shang cosmology. Such texts will become important later in this study, as sources concerning how Warring States and Han people reconstructed, transformed, and reinvented Shang cosmology, rather than as evidence of Shang cosmology itself. Shang oracle inscriptions – divination records engraved on tortoise shells or cattle bones – will be the sole literary source.[8] The analysis of the Western Zhou will include two new sources: the long inscriptions on Western Zhou bronze vessels,[9] and transmitted texts that derive from original Western Zhou writings, found in the *Book of Documents* (*Shangshu*) 尚書 and the *Book of Poetry* (*Shijing*) 詩經.[10]

8 For the transcription of oracle bone inscriptions I rely primarily on Yao Xiaosui (1989), the translation from Chinese to English being my own. For my translations, I have taken into consideration the principles laid out by David Keightley (1978b), as well as the debate between Qiu Xigui (1989) and David Nivison (1989) over whether the inscriptions should be interpreted as questions.

9 For a detailed study of the nature and structure of the Western Zhou bronze inscriptions, see Shaughnessy, 1991. Shaughnessy treats these inscriptions as the primary source of Western Zhou history – the first conscious attempts at writing history, written for later generations (1991, pp. 1, 181–2). Lothar von Falkenhausen cautions that bronze inscriptions are essentially religious documents, cast on ritual objects and used to communicate with the ancestors. See Falkenhausen, 1993, pp. 145–52.

10 Shaughnessy has compared the structure and content of some transmitted texts to the bronze inscriptions, and finds striking similarities between the two. See Shaughnessy, 1991, pp. 73–6.

Sifang (Four Quarters)

Political Geography: Fang *versus* Wo

Since *Sifang* means four *fang*, the term *fang* must first be defined. Most scholars translate *Sifang* as "four directions." Paul Wheatley translates *Sifang* as "four quarters," seeing the *fang* as cardinal orientations.[11] Sarah Allan, however, insists that *fang* refers to a square space or cube rather than to a linear direction, since the graphic *fang* 方 is made up of an element of *ren* 人 holding a tool 工, likely representing a carpenter holding a tool used in making squares.[12]

Moving from Allan's graphic analysis of *fang* to the context in which the graphic was used, I would argue that in Shang oracle bone inscriptions *fang* is primarily a concept of political geography. *Fang* most often describes alien polities, referred to either as "x *fang*," indicating a specific polity, or simply as "*fang*" or "many *fang*" (*duo fang* 多方), as opposed to "us" (*wo* 我) – that is, the Shang state. Yao Xiaosui identifies forty polities in oracle bone inscriptions whose names consist of an individual name preceding the word *fang*, thus appearing as "x *fang*" (e.g., Gongfang, Renfang). Shima Kunio has found fifty-three such *fang*. Including some polities whose names did not include *fang*, Chen Mengjia lists forty-seven *fang* polities, and Zhong Bosheng lists eighty-four.[13] In this political context, Keightley's translation of *fang* as "side, border, country, or region" best conveys the primary meaning of *fang* as a boundary marker of the Shang world, differentiating the Shang from all the alien, hostile, or unknown others.[14]

The political center of the Shang state was surrounded and thus defined by these alien polities. The domain of the Shang was composed of an "inner area" and an "outer area." The former was called the "Zhong Shang 中商" or "*zhong tu* 中土" – the Central Shang or central land, including the ancestral capital, the present capital, and the royal hunting area. The "outer area" included four lands (*si tu* 四土) – the Eastern, Northern, Western, and Southern lands – and all the Shang lords.[15] The center and its outer domain were further defined by many

11 Wheatley, 1971, pp. 423–7. 12 Allen, 1991, pp. 75–7.

13 See Yao Xiaosui, 1989, vol. 3, pp. 1203–4; Shima Kunio, 1958, pp. 384–5; Chen Mengjia, 1956, pp. 269–312; Zhong Bosheng, 1989, p. 169.

14 Keightley, 1999, p. 269.

15 Some scholars have identified the four lands with the four *fang*, others have tried to distinguish them. Sarah Allan has made the distinction that the four lands were real lands, while the *fang* were spiritual lands. See Allan, 1991, pp. 83–4. While I agree that the two concepts should be distinguished, I prefer to make two alternative distinctions. First, the distinction between the four lands and four *fang* was primarily one of

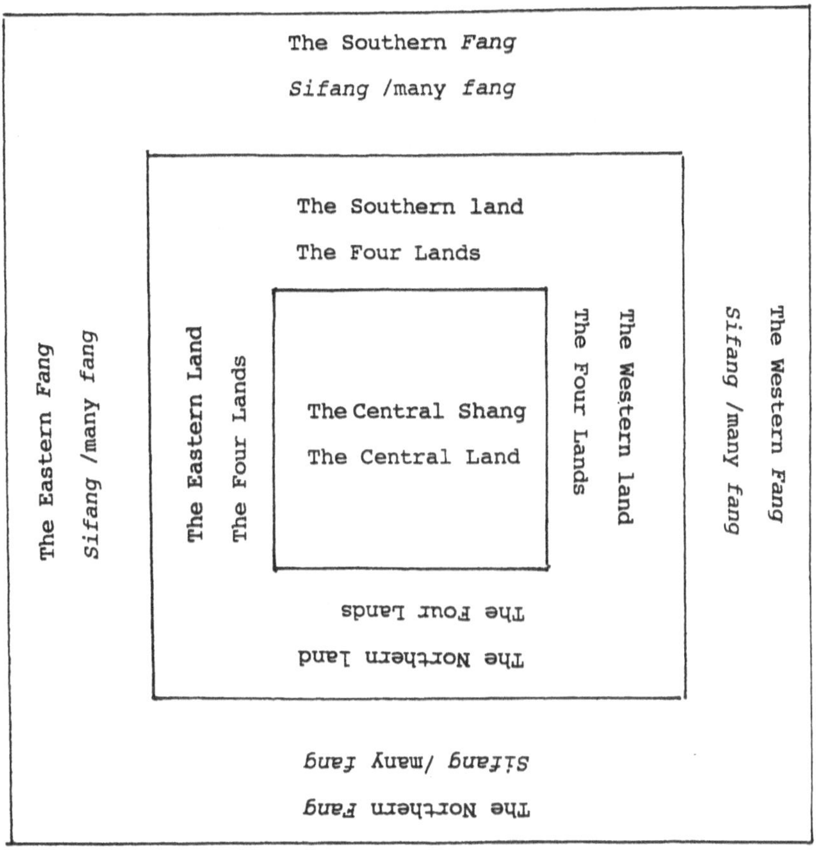

Figure 2.1. The Shang conception of political geography.

fang who lived outside of the four lands, and who were most often alien to the Shang.[16] In this context, *fang* represented "otherness" and "outerness" in contrast to the centrality of "us"; the centrality of a homogeneous "us" was defined in turn by contrast to the heterogeneous "others" – *fang* (Fig. 2.1).

political geography – the lands were domains of Shang itself, while *fang* were alien polities outside of the Shang lands. Secondly, as I shall demonstrate later in this chapter, *fang* in *Sifang* refers to a cosmological structure that embodies real lands as well as spiritual ones, structuring political as well as ritual relationships.

16 The political and economic structure of the Shang state as well as its relationship to other polities has been extensively described and debated by David Keightley and Kwang-chih Chang. See Keightley, 1983, pp. 523–65; and Kwang-chih Chang, 1980a, pp. 210–60.

The boundary between the others – the alien polities – and the center – the Shang – was not marked by fixed territories, but was constantly defined and redefined through dynamic interactions. According to Shang oracle inscriptions, these interactions included: (1) invading or raiding each other's territories and livestock (*qin* 侵, *fa* 伐, *zheng* 征, etc.); (2) forming alliances or dominant-subordinate relationships (*hu* 呼, *ling* 令, *ce* 冊, etc.); (3) giving blessing to or cursing one another (*sui* 祟, *shou you* 受佑, *huo* 禍, etc.); and (4) trading and giving tribute (*lai* 來), where "x *fang*" sent in shells or other goods. According to its own account in oracle bone inscriptions, the Shang usually had the upper hand in these interactions, and its domination, while challenged, was not undermined until the end of the Shang period.[17]

These constant interactions between the many *fang* and the Shang created extremely fluid relationships. A *fang* could be absorbed into the Shang state, as in the case of Yangfang 羊方, or a lord of the Shang could rebel and become a *fang*, as in the case of Yufang 盂方. Under some circumstances, the *fang* and the Shang's own lords were interchangeable.[18] Yang Nansheng has demonstrated with rich evidence that when a *fang* was conquered by the Shang, its name was changed from "x *fang*" to "x *hou* 侯 (lord)," indicating the status of a subject, and when such a lord rebelled against the Shang, it again became "x *fang*" and was subjected to military reconquest.[19] As fluid and shifting as such relationships were, what differentiated "us" at the center from the "others" at the periphery can clearly be seen in the Shang king's concern for harvests; he divined for the harvests of his capital at Shang, his four lands, all his domain (*wo* 我), and his lords, but did not divine for the harvest of a specific *fang* or "many *fang*."[20]

Cosmological Structure: Sifang

While *fang* represents the others and the periphery in political geography, the meaning of *Sifang* (four *fang*) extends to a more comprehensive spatial structure of cosmology in which political geography is included. When combined with the number four in *Sifang*, *fang* acquired a cosmological meaning that went beyond the concept of *fang* as a polity. Rather than indicating simply a periphery that serves to define a

17 Detailed studies of relations between Shang and the many *fang* include Shima Kunio, 1958; Hu Houxuan, 1944a; and Chen Mengjia, 1956.

18 Kwang-chih Chang, 1980a, pp. 248–59.

19 Yang Nansheng, 1983, pp. 132–3.

20 Chen Mengjia, 1956, p. 639. Chen's conclusion is further supported by my research in oracle bone inscriptions collected in Yao Xiaosui's *Leizuan*. I found no case of harvest divination under entries for any specific *fang*, or under "many *fang*."

political center, *Sifang* also represents the four cardinal directions, further defining the center in ritual and cosmological terms.

In oracle bone inscriptions, *Sifang* as such a cosmological structure was expressed interchangeably in several ways – as the four *fang*, as Eastern, Western, Southern, and Northern *fang*, as the four directions (East, West, South, and North), or simply as *fang*. For example, in a few pieces of inscription in which the four *fang* are described in detail, each *fang* has its own name and is associated with a wind that is also named. One of the most detailed records concerning the four *fang* and the wind is translated as follows:

東方曰析風曰劦
南方曰因風曰凯
西方曰𣏟風曰彝
〔北方曰〕勹風曰𠬝
The Eastern *fang* is called 析, its wind is called 劦.
The Southern *fang* is called 因, its wind is called 凯.
The Western *fang* is called 𣏟, its wind is called 彝.
The Northern *fang* is called 勹, its wind is called 𠬝.

(*Heji*, 14294)

In analyzing this inscription in conjunction with the few others on *Sifang* and the wind, scholars generally agree that *Sifang* as the four cardinal directions was associated with the four winds as well as with the gods of the winds.[21] In other inscriptions, such a complete account of the four *fang* was often abbreviated as *Sifang*; as the East, West, North, and South; or simply as *fang*. The best example of such an abbreviation is the *di*-sacrifice (禘) that was performed almost exclusively to *Sifang*. It could be elaborately described with the names of each *fang* and the names of the winds associated with each *fang*, as quoted later in *Heji*, 14295, or simply denoted as "*di*-sacrifice to the North (*di bei* 禘北)," "*di*-sacrifice to the South (*di nan* 禘南)," "*di*-sacrifice to the West (*di xi* 禘西)" (*Heji*, 34154), or even more simply as "*di*-sacrifice to *fang* (*di yu fang* 禘于方)."

The same word *fang*, therefore, could either refer to alien polities or serve as an abbreviation for *Sifang*. This lack of distinction between the two uses demonstrates the overlap of the two concepts, both referring to the boundary of the Shang – the political boundary marked by the many *fang* polities and the cosmological boundary marked by powers and gods living beyond the Shang control. As shown in the following discussion, *Sifang* as four directions and a classification struc-

21 The major studies on *Sifang* and the four winds include Hu Houxuan, 1944b; 1956; Akatsuka Kiyoshi, 1977, pp. 415–43, Ikeda Suetoshi, 1981, pp. 122–37; Li Xueqin, 1985; Allan, 1991, pp. 74–98; and Feng Shi, 1994.

ture included alien people living on the periphery and the gods of *Sifang*, who had spiritual influence over the actions of alien polities in the same way that they influenced rain and wind coming from the four directions.

Expressed in these interchangeable terms, *Sifang* as a cosmological structure classified all forces of the universe, including spirits, beings, and natural powers, as well as alien polities, on the basis of the four cardinal directions. For the Shang people, all these forces were commanded by the high god *Di* 帝, who used them to determine the well-being of the Shang. Most scholars agree that *Di* was the supreme, abstract deity in Shang theology, controlling natural powers as well as spirits.[22] There has been a prolonged debate, however, regarding the origin and evolution of this concept. Some scholars, such as Akatsuka Kiyoshi and Ikeda Suetoshi, contend that *Di* originated from the ancestors of Shang and later developed into the abstract high god of the universe.[23] Chen Mengjia, Ito Michiharu, and Hayashi Mino, by contrast, believe that *Di* in early Shang was an abstract high god controlling natural forces as well as ancestors. Ito and Hayashi further suggest that such an abstract concept of *Di* was used to absorb and integrate the gods and spirits of many divergent groups that Shang contacted, and that it was only toward the end of the Shang period that *Di* was reduced to representing only Shang's ancestors.[24] Robert Eno recently revisited the subject by arguing that the Shang concept of *Di* was a generic or corporate term derived from its earlier function, that of denoting the ancesters of a lineage, and that the term later came to include the natural spirits and gods of conquered groups.[25] The two opposing hypotheses – that *Di* was originally an abstract singular high god that developed into an ancestral god, and that *Di* was originally a term for ancestors and later assumed the position of high deity – both propose that *Di*, as an abstract high god, evolved through the long process of Shang's interaction with alien groups, and that it functioned as a concept incorporating spirits of nature and alien groups in the Shang theology. This function of the high god *Di* in absorbing deities of other groups was most important, regardless of whether the change happened in the early or late Shang.

This observation is of particular value for my study on Shang cosmology. The following analysis of *Sifang* cosmology elaborates on

22 The most extensive studies of *Di* as the high god include Hu Houxuan, 1959; Chen Mengjia, 1956, pp. 561–82; Shima Kunio, 1958; Akatsuka Kiyoshi, 1977, pp. 471–610; Keightley, 1978a; and Ikeda Suetoshi, 1981, pp. 25–63.

23 Akatsuka, 1977, pp. 471–610; Ikeda, 1981, pp. 25–63.

24 Chen Mengjia, 1956, pp. 561–82; Hayashi Mino, 1970; and Ito Michiharu, 1975, pp. 45, 55–79.

25 Eno, 1990.

this point to show that it was precisely in the conjunction of the Shang's high god *Di* and the natural forces and alien spirits that *Sifang* cosmology became an indispensable medium. (It is in this sense that I translate *Di* as a singular deity, aware of the possibility that during the early Shang period, *Di* may have referred only to ancestors.) In Shang theology, the high god *Di* and numerous natural forces and alien spirits were connected in the cosmological structure of *Sifang*. These forces and spirits were not randomly sent by the unpredictable will of the high god, but rather sent through the *Sifang* structure, ordered in time and space. As will be demonstrated, rain, clouds, spirits, harvest, disaster, illness, and raids from alien groups did not just happen to the Shang; they arrived from the four *fang* upon the center – Shang – and thus the direction from which they came became an important subject for divination. Therefore, *Sifang* became the conception through which the Shang perceived the will of *Di*, and the forces and spirits that *Di* sent through *Sifang* became the mediators between *Di* and the Shang people.

A wide range of forces was conceptualized in terms of such mediators. They can be grouped into four major categories: (1) natural phenomena; (2) ill fortune; (3) visits and tributes, as well as raids coming from the outside; and (4) blessings, rain, and harvest.

Natural phenomena were perceived as the messengers or ministers of the high god, ascending and descending through *Sifang*, passing *Di*'s messages, and running his errands. These phenomena included rain, clouds, wind, thunder, and rainbows, among others. For instance, the wind was considered the minister of the high god, called *Di shi Feng* 帝史風.[26] Rainbows and clouds were considered forerunners of disaster, coming from *Sifang*:[27]

王固曰：有祟
八日庚戌有各雲自東[illegible]母
昃亦有出虹自北飲于河

The king read the crack and said: "there will be evil fortune." On the eighth day *gengxu*, there were dark clouds coming from the East [illegible]Mother, and that afternoon there was also a rainbow coming out from the North and drinking from the River.

(*Heji*, 10405b and 10406b)

And rain, of crucial importance for the harvest, was perceived as another divine force descending through *Sifang*:

26 Hu Houxuan, 1959. Oracle bone inscriptions are cited primarily from the following collections by abbreviation: *Heji*: Guo Moruo, 1982; and *Tun*: Xiao Nan, 1980–83.
27 Hu Houxuan, 1944d, pp. 19b-20b.

癸卯卜今日雨
其自東來雨
其自南來雨
其自西來雨
其自北來雨

Crack on day *guimao*: It will rain today.
There may be rain coming from the East.
There may be rain coming from the South.
There may be rain coming from the West.
There may be rain coming from the North.

(*Heji*, 12870)

The second category of mediators sent by *Di* consisted of ill fortune (*sui* 祟), which came through *Sifang* and brought with it disasters (*huo* 禍) and curses (*tuo* 壱). For example:

癸酉貞：旬有祟自東有來禍
癸酉貞：旬有祟自南有來禍

Divining on day *guiyou*: In this ten-day week, there will be ill fortune coming from the East and bringing disaster.
Divining on day *guiyou*: In this ten-day week, there will be ill fortune coming from the South and bringing disaster.

(*Tun*, 2446)

.....惟西方壱我

..... The Western *Fang* will curse us.

(*Heji*, 33094)

Thus, for the Shang, disasters were brought about by "ill fortune" from *Sifang*, or by one of the four *fang* directly. Raids and invasions of Shang territories by alien states were a specific kind of ill fortune, also coming through *Sifang*. Two characters, *gu* 娃 (*kagx*) and *jian* 艱 (*kran*) are terms used to depict distress or distressing news of this kind.[28] For example:

王固曰：有祟。其有來娃
迄至七日乙巳允有來娃自西
𢀛友角告曰：舌方出侵我示䊮田七十五人

The king read the crack and said: "there will be ill fortune, which is the coming calamities 娃." Up to the seventh day *yisi*, indeed the calamities 娃 came from the West. 𢀛 You Jiao reported that Gongfang came to invade our *shi* 䊮 field, [captured] seventy-five of [our] people.

(*Heji*, 6057a)

A third category of mediators consisted of visitors and tribute, as well as harm from polities outside of the Shang, again coming through the

28 Schuessler, 1987, pp. 290–1; Xu Zhongshu, 1989.

Sifang structure. These outside impacts are represented by the term "coming" (*lai* 來). One significant meaning of "coming" indicates visits paid by outside groups and the tribute they brought to Shang in material or symbolic form. Such tribute included turtles (*lai gui* 來龜; *Heji*, 7076a), cattle (*lai chu* 來芻; *Heji*, 102), animal teeth (*lai chi* 來齒; *Ying*, 886a), horses (*lai ma* 來馬; *Heji*, 9173), oxen (*lai niu* 來牛; *Heji*, 9179), women (*lai nü* 來女; *Heji*, 668), the captives from Qiangfang used for sacrifices (*lai Qiang* 來羌; *Heji*, 226a), dogs (*lai quan* 來犬; *Heji*, 21562), and timber (*lai mu* 來木; *Huai*, 1629). These tributes, which we would consider political or economic activities, were considered by the Shang as forces fulfilling *Di*'s will.

Besides the large number of records of the "coming" of specific kinds of tribute, the term denotes the coming of outside forces, impact from the outside. Such forces could include rain, disasters, spirits, and raids or visits by other polities. The majority of divination records on whether or not there will be a "coming," however, did not indicate what exactly was coming. Some such records give a precise number, such as "a coming of thirty [x] from Dian 奠來三十" (*Heji*, 9613b), or "a coming of forty [x] from Jian 戔來四十" (*Heji*, 438b). These records can be regarded as documentation of tribute, as they were most likely abbreviations of records such as "a coming of fifty Qiang [captives] 來羌五十" (*Heji*, 226a), or "a coming of five white horses 來白馬五" (*Heji*, 9177a). But the largest number of divinations about whether there would be a "coming" did not give either the object or a specific number. These could be interpreted in a narrow sense as records about visits and tribute from outside groups, or in a broad sense as records of outside forces or impacts. In either case, however, the visits or forces descended upon the Shang at the center through the *Sifang* structure from the outside. The following four individual cases are divinations about a "coming" from *Sifang*:

王固曰：其自東來
The king read the crack and said: "There will be [x] coming from the East."
(*Heji*, 914b)

有來自南以龜
There will be turtles coming from the South.
(*Heji*, 7076a)

丙申卜古貞：有來自西
Crack on day *bingshen*, Gu divining: There will be [x] coming from the West.
(*Heji*, 7112)

. . . 韋貞：有來自北
. . . Wei divining: There will be [x] coming from the North.
(*Heji*, 7121a)

The fourth category – blessings (*shou you* 受侑), rain, and harvests (*shou nian* 受年 and *shou he* 受禾) – was also seen as sent by *Di* to the Shang through *Sifang*. The Shang king prayed for blessings and good harvests to *Di*, to the rivers, to the mountains, and to ancestors, as well as to *Sifang*. *Sifang* was not only an object of ritual sacrifice and prayer, as extensively demonstrated in the following section on ritual, but also a structure through which blessings, rain, and harvest descended on Shang. For example, the harvest came from *Sifang*:

乙未卜貞：今歲受年
不受年
南受年
東受年 . . .
Crack on day *yiwei*, divining: This year will receive harvest.
It will not receive harvest.
The South shall receive harvest.
The East shall receive harvest . . .

(*Heji*, 36976)

The blessing also descends through the *Sifang* structure:

卜佑四方
Crack-making: Blessing with regard to *Sifang*.

(*Tun*, 3661)

Such evidence demonstrates that, in the Shang world, *Sifang* was a structure classifying all forces in the universe, the channel through which these forces traveled expressing *Di*'s will, affecting the well-being of the Shang. *Sifang* represented the Shang conception of the universe, in which the highest deity, *Di*, integrated and controlled the myriad natural forces and spirits.

Sifang as a classification structure was directly linked with Shang ritual and political action. Classifying all forces and spirits by means of this structure, the Shang king not only predicted what was coming to Shang, but also distinguished which force had sent it and where it came from. With this knowledge, he could influence such forces through ritual sacrifice. *Sifang* thereby evolved from a structure of classification to a structure for ritual activities. As I shall argue in the next section, it was through this *Sifang* structure that the Shang king communicated, interacted, and negotiated with the high god *Di* and the world of divine powers.

Ritual Structure: Di *and* Fang

As a structure of classification of myriad forces – consolidating Shang's high god *Di* with the natural forces and alien spirits and mediating between the human world and the divine world – *Sifang* cosmology became a primary structure of Shang ritual action. Shang rituals constituted the king's monopolized access to divine powers, and that access provided a foundation for political domination. The following evidence shows how *Sifang* functioned as the primary structure for ritual; it was through *Sifang* that the Shang kings inquired about, made offerings to, prayed to, and negotiated with divine powers.

Rituals performed to *Sifang* can be divided into two major groups according to their purpose – appealing rituals, performed for divine assistance; and alleviation rituals, performed for abating disasters and sufferings.

The appealing rituals requested rain, harvest, and blessing (*you* 佑) in the form of divine assistance in military campaigns or the well-being of the Shang in general rituals. Although these were sacrificial rituals communicating with *Di* and the divine world, they were often made specifically to *Sifang*. The appealing sacrifices took two primary forms – burning sacrifices (*liao* 燎) and *di*-sacrifices (禘). Burning sacrifices were made to a wide range of deities, including ancestors, and to natural forces such as mountains, rivers, clouds, and so on. Besides the ancestors of the Shang king, *Sifang* was the most frequent object of burning sacrifices according to Yao Xiaoxui's concordance. *Di*-sacrifices, however, were made almost exclusively to *Sifang*, and the most extensive records on *Sifang* dealt with the performance of *di*-sacrifices:[29]

辛亥內貞：今一月帝令雨。四日甲寅夕。。。一二三四
辛亥卜內貞：今一月〔帝〕不其令雨。一三三四
辛亥卜內貞：禘于北，方曰勹，風曰殴。𠦪〔年〕。一二三四
辛亥卜內貞：禘于南，方曰兂，風〔曰〕乚。𠦪年。一月。一二三四
貞：禘于東，方曰析，風曰劦。𠦪年。一二三〔四〕
貞：禘于西，方曰彝，風曰[illegible]。𠦪年。一二三四

On day *xinhai*, Nei divining: This first month, *Di* [the high god] will command rain. On the evening of the fourth day, *jiayin*, [it indeed rained.] 1, 2, 3, 4.

Crack on day *xinhai*, Nei divining: This first month, [*Di*] will not command rain. 1,2,3,4.

Crack on day *xinhai*, Nei divining: Offer the *di*-sacrifice to the North, its *fang* is called 勹, its wind is called 殴. Pray for [harvest.] 1, 2, 3, 4.

Crack on day *xinhai*, Nei divining: Offer the *di*-sacrifice to the South, its *fang* is called 兂, its wind [is called] 乚. Pray for harvest. First month. 1, 2, 3, 4.

Divining: Offer the *di*-sacrifice to the East, its *fang* is called 析, its wind is called 劦. Pray for harvest. 1, 2, 3, [4.]

29 The transcription of this record is adopted from Feng Shi, 1994.

Divining: Offer the *di*-sacrifice to the West, its *fang* is called 𠦪, [its wind] is called 𠦪. Pray for harvest. 1, 2, 3, 4.

(*Heji*, 14295)

The elaborate divination and *di*-sacrificial ritual for blessing or harvest was sometimes abbreviated into a code-like fragment, with *Sifang* abbreviated as *fang*:

其𠦪年于方受年
Pray for harvest to *fang*, receive harvest

(*Heji*, 28244)

Rituals praying for rain were also performed to *Sifang*, for example:

甲子卜：其𠦪雨于東方
Crack on day *jiazi*: Pray for rain to the Eastern *fang*.

(*Heji*, 30173)

Alleviation rituals, the second kind of ritual made to *Sifang*, attempted to alleviate calamities and difficulties that the Shang currently suffered, such as illness, disasters of various kinds, and damaging wind. These were sacrificial rites using animals and human victims (often those captured from a group called *Qiang*) as offerings. In these rituals the Shang king asked for the pacifying of illness or wind, and prayed for freedom from disaster:

壬辰卜：其寧疾于四方。三羌九犬
Crack on day *renchen*: It will pacify the illness at *Sifang*. [The sacrifice used] three Qiang (victims), and nine dogs.

(*Tun*, 1059)

A very similar record to the one above on pacifying wind addressed *Sifang* simply as *fang*:

癸未卜：其寧風于方有雨
Crack on day *guiwei*: It will pacify the wind at *fang* and rain.

(*Heji*, 30260)

This evidence demonstrates that frequent and elaborate rituals were a means of communicating with the divine world, and that the Shang king depended on the *Sifang* structure for such ritual communications. None of the sacrifices was offered directly to the high god *Di*. For the Shang, the high god did not interact directly with human beings, but rather sent forces in the universe as his mediator through the *Sifang* structure. Sacrificial rites, therefore, had to be directed to the mediators and to *Sifang* – where *Di*, the mediators, and the human world intersected – rather than to the high god directly.

Because *Sifang* was the primary structure for ritual communication with the divine world, the highest of the Shang sacrificial rites, the *di*-sacrifice, was made primarily to *Sifang*. All but three of the seventy-five *di*-sacrifices recorded in Yao Xiaosui's *Leizuan* were made to *Sifang*, in terms such as "*di yu fang* 禘于方" (or "*di yu* + North, South, East, or West"), "*fang di* 方禘," "*di yu wu* 禘于巫," or "*wu di* 巫禘." All the *wu* in the context of di-sacrifice are written as [glyph]. Chen Mengjia, David S. Nivison, and Sarah Allan have elaborately argued that *wu* in this case represents the four cardinal directions and means *fang* or *Sifang*.[30] Furthermore, the character *di* 禘, written in oracle inscriptions as [glyph], stands for *di*-sacrifice, the high god (*Shang Di*), and lower gods (*xia di,* composed of royal ancestors).[31] *Di* [glyph], like *wu* [glyph], visually suggests cardinal directions crossing a central point. Allan has pointed out that when *di*-sacrifice is used as a verb, a square □ sometimes substitutes for the element [glyph]: [glyph], suggesting that this square symbolizes *Sifang*.[32] Although explaining the meaning of a word by analyzing the component of the graphic is a linguistically challengable method,[33] the ritual context seems to accord with Allan's graphic analysis. My research shows that *di*-sacrifice communicating to the high god has to be made through the structure of *Sifang*; this could explain why *di*-sacrifice and the high god *Di* itself both had a *Sifang* structure as their graphic component.

Centrality

It is in this capacity – connecting human beings to the world of the gods – that *Sifang* as four cardinal directions points to a center. It was at this center that political power was rooted, access to the world of the gods was monopolized, and divine knowledge – of the cosmos and of the world of gods and spirits – was achieved through divination and ritual. It was also at this center that such divine knowledge and political domination came together. Since the high god was the commander of all forces of the universe, monopolizing access to the world of *Di* meant winning divine assistance and manipulating cosmological forces, as well as dominating political power. The Shang kings achieved this monopoly by way of ancestor worship.

The royal ancestral line in the world of the Shang was seen as the

30 See Chen Mengjia, 1956, p. 579; Nivison, 1987; and Allan, 1991, p. 77.

31 Xu Zhongshu, 1989. 32 Allan, 1991, p. 78.

33 The traditional method of viewing Chinese writing as a form of "ideograph" has been widely challenged as being linguistically unsound. For a recent summary and response to such challenges, see Hansen, 1993.

cosmological center and as the juncture of the four *fang*. Influenced by Marcel Eliade's theory of the symbolism of the center as the zone of the sacred, some scholars have tried to locate this center geographically in China by looking for a sacred mountain.[34] Others, also conceiving the center as a specific place, have tried to locate it at the capital of the kings, an effort best represented by Wheatley's monumental work.[35] But, since the Shang kings moved their capitals frequently, the geographic location of the central capital remains a problem. David Keightley accordingly denies the existence of such a central capital in Shang times. He suggests that the Shang king was peripatetic, displaying his power by frequently traveling, hunting, and inspecting his domains, bringing along with him his court and diviners.[36] Another hypothesis, first suggested by Dong Zuobin, is that it was the ancestral capital, rather than the political capitals of the kings, that was the center of the Shang world.[37] K. C. Chang has further developed this hypothesis, revealing that the Xia, Shang, and Zhou Dynasties regularly moved their administrative capitals while retaining the dynasty's ancestral capital as the sacred ritual center. While the sacred capital remained fixed, the secular capitals moved constantly as part of the search for copper and tin mines for bronze production. Bronze vessels served as the medium for communication between Heaven and Earth, and possessing bronze meant monopolizing access to the divine knowledge that was the basis of political authority.[38]

Whether it is Keightley's hypothesis that the Shang king was peripatetic, or Chang's hypothesis that the ancestral temple was a fixed ritual center, the center of Shang political power resided in the king's body and his ritual connection to his ancestors. This centrality and connection to the king's ancestral line are further supported by Nancy Price's recent comparative study of the "center" in early civilizations. Challenging the "Euro-American" model of the "center" as a fixed place – a "ceremonial center" or a capital that defines a complex society – Price has proposed a model of a "moving" center, in which power is not necessarily fixed at a particular place but rather is mobile. Using this model to explain Shang history, the centrality of the Shang king is defined not by a specific fixed locality but by means of orientation to the "abstract framework" defined by *Sifang*.[39]

34 For discussions of Eliade's symbolism of the center and its application to the study of Chinese tradition, see Allan, 1991, pp. 98–101.
35 Wheatley, 1971, pp. 428–51. For discussions of the Shang capitals as the center of the universe, see Chen Mengjia, 1956.
36 Keightley, 1983. 37 Dong Zuobin, 1945, vol. 9, p. 12.
38 Kwang-chih Chang, 1984. 39 Price, 1995.

While the location and mobility of the ritual-political center of the Shang remain under debate, the centrality of the king's body and his ancestral line was clearly defined in the cosmology of *Sifang* and Shang theology. This centrality was fully manifested, as an "abstract framework," in the political and material cultures of the time. Politically, ancestor worship constituted the basis for political domination. According to Hu Houxuan's research, ancestors of the Shang king ascended to the world of *Di* to be *Di*'s company (*bin Di* 賓帝). They had power almost comparable to that of *Di* himself. As mobile as the living king, the deceased ancestors were able to ascend and descend freely, to bring down blessings or ill fortune on Shang, ruin harvests, spoil rain, cause illness and disaster, order alien states to raid, or assist the Shang king. They were even given the title of "*Di*" after their death, addressed as the lower god (*Xia Di* 下帝) as opposed to the high god (*Shang Di* 上帝).[40]

Yet the ultimate distinction between the ancestors and *Di* was that *Di* did not receive prayers from human beings directly; all prayers had to be made to the ancestors, who transmitted the messages to the high god.[41] This distinction pinpoints the political function of ancestor worship. Only the ancestors of the king could pass messages directly to the high god, and only the king could communicate with the ancestors. The king's authority and ability to perform sacrifices and to pray to his ancestors thus became the source of his political power and the legitimization of the Shang state. As Keightley has best summarized: "The king's ability to determine through divination, and influence through prayer and sacrifice, the will of the ancestral spirits legitimized the concentration of political power in his person."[42]

Culturally, the centrality of royal ancestor worship was symbolized by material constructions, particularly in the *Sifang*-center structure called the *ya* 亞-shape by many scholars. Evidence of such material constructions can be grouped into three categories:

1. Textual data: bronze inscriptions and later texts suggesting that Shang and Zhou ancestral temples were constructed using the *Sifang*-center structure.
2. Archaeological data: the Shang royal tombs and their wooden chambers that employed the *Sifang*-center structure.
3. Epigraphic data: Shang and Zhou inscriptions and bronze emblems in the *Sifang*-center *ya*-shape.

40 As mentioned earlier, Eno believes that the very concept of an abstract high god *Di* actually derived from an early usage of *Di* as the generic term for deceased kings. See Eno, 1990.

41 Hu Houxuan, 1959. 42 Keightley, 1978a, pp. 212–13.

The idea that ancestral temples of the Bronze Age were built using the *ya*-shaped *Sifang*-center structure has been a dominant theory ever since the *Bogutu*, one of the earliest antiquarian catalogues compiled in 1107–1125 during the Song Dynasty. This theory is supported by a body of literary data that has been thoroughly explored. Relying on Zhou bronze inscriptions and later texts, for example, Wang Guowei has reconstructed the ancestral temples of early China as *ya*-shaped, with a square central chamber and four additional chambers attached at the four sides. He further concludes that such *ya*-shaped ancestral temples constitued what was later called a "bright hall" (*Ming tang* 明堂) – the common structure for ancient ritual halls in later literature.[43] Chen Mengjia has further explored the ritual function of this structure as recorded in Shang and Zhou inscriptions. He reveals that the ancestral altar was located in the central space, called *zhong zong* 中宗, *ya zong* 亞宗, and so on, while the four chambers in the four directions were places where sacrificial rites were performed.[44] Zhu Fenghan has recently interpreted the frequent appearances of the word *ya* in Shang oracle inscription as referring to the ancestral temples.[45] Keightley suggests that Shang architectural units were marked in terms of the four cardinal directions, such as "the north temple 北宗," "the north altar stand 北示," "the south camber 南室," "the east hall 東寢," "the west hall 西寢," "the south gate 南門." While the overall architectural plan of these units is unknown, the great attention given to the directionality suggests their religious function.[46] This speculation, however, awaits further archaeological evidence, since no Shang ancestral temple has yet been found, and the earliest evidence of a *ya*-shaped ritual construction is dated at the end of the Former Han, more than a thousand years later than the Shang.

Without a Shang ancestral temple excavated to prove the hypothesis of a *ya*-shaped architectural design, Wu Hung has reconstructed the Bronze Age temple based on three excavated temple-palace foundations, two dated in late Xia (sixteenth century B.C.) and one in early Western Zhou (eleventh-tenth centuries B.C.). Although none of these foundations was an ancestral temple of the Shang, nor were any of them in a *ya*-shape, the structures symbolized the same centrality of ancestor worship by the king: "the corridors created discontinuity in space by separating the 'inner' from the 'outer,' and the enclosed open courtyard then became a relatively independent space with its focus in the central hall." Deep inside the temple compound were housed the bronze

43 Wang Guowei, (1921) 1984, vol. 3, pp. 1–11.
44 Chen Mengjia, 1956, pp. 473–82. 45 Zhu Fenghan, 1990, pp. 169–71.
46 Keightley, forthcoming, ch. 6, n. 27, 28.

vessels, the means to communicate with the "invisible spirits of ancestors." In this structure, not a central point but the temple's entire complex became the "metaphor" for ancestral religion itself.[47]

The second body of evidence, archaeological data, is best represented by Gao Quxun's paper on the excavations of the royal cemetery near the last capital of Shang. These excavations uncovered twelve royal tombs, half of which have a *ya*-shaped wooden chamber, either with or without a *ya*-shaped earthen pit, and long ramps leading from the ground to the bottom of the tomb at either four or two sides. Gao observes that because the *ya*-shaped wooden chamber required a tremendous amount of labor and material and was difficult to construct, it must have had a particular meaning, which he believes was to symbolize the ancestral temples of the Bronze Age, which he assumes were in the *ya*-shape as reconstructed by Wang Guowei.[48]

Reports of these excavations also reveal that many human sacrifices had occurred in these graves, mostly in the four side chambers and in the ramps. The best example is tomb 1001, which has both the earthen pit and the wooden chamber in a *ya*-shape, and four ramps leading to four directions. The opening of the pit is nineteen meters long, south to north, and fourteen meters wide (see Fig. 2.2). Over ninety human victims were unearthed within the tomb, and they were distributed in three different locations: (1) nine sacrificial burials at the bottom of the chamber, one at the center and two at each of the four corners (see Fig. 2.3); (2) twelve human victims distributed at the top and sides of the chamber; (3) victims with body and head separated – sixty-one headless bodies and seventy-three heads found – buried in the four ramps, with a few in pits outside the chambers. Most victims of this group were buried in the southern ramp. Bodies and heads were buried in lines oriented to the four directions (see Fig. 2.4). Another example is tomb 1550, square at the opening and *ya*-shaped at the bottom, with most parts of the tomb pit damaged. Four sacrificial burials and 243 human skulls were found in the pit, including 235 in the northern ramps and 8 in the southern ramp. These skulls were buried in east-west lines, with the top up and facing the center.[49] These human sacrifices, then, were clearly arranged according to the four directions, like the tomb structure itself, and were oriented toward the center, where the Shang royal ancestor was buried.

The ritual and political function of the third group of materials, *ya*-shaped inscriptions, is still highly hypothetical, although much discussed

47 Wu Hung, 1995, pp. 79–88. 48 Gao Quxun, 1969.
49 Liang Siyong and Gao Quxun, 1962, 1974, 1976.

Figure 2.2. Grave pit of Shang royal tomb no. 1001 (Liang Siyong and Gao Quxun, *Houjiazhuang*, Taipei: Institute of History and Philology, 1962, vol. 2, pl. 3b. Courtesy of the Institute of History and Philology, Academia Sinica).

(see Fig. 2.5). The generally accepted interpretation since the Song Dynasty has been that they symbolize the *ya*-shaped ancestral temple. Some modern scholars have challenged this interpretation with various new hypotheses.[50] Among them, Liu Jie suggests that *ya* means lineage and that the term "many *ya*" was used in contrast to "many *fang*"; the former were "our own lineage," that of the Shang king, and the latter were the lineage of the others, the alien groups.[51] Two recent studies on Shang clan-signs by Zhu Fenghan and Noel Barnard suggest that the *ya*-shape is a component frequently used in the combined Shang clan-signs. Zhu Fenghan stresses that the combined clan-signs, including the ones in the *ya*-shape, signal the blood ties of common ancestors that bound various branches of the same clan.[52] Barnard actually discovered that some of the *ya*-shaped emblems enclosed ancestral figures of Shang – *ri*

50 For a summary and evaluation of these hypotheses, see Kwang-chih Chang, 1990.
51 Liu Jie, *Zhongguo gudai zongzu yizhi Shilun* 中國古代宗族移植史論, pp. 14–15, as quoted in Zhou Fagao, 1975, no. 14.1833, pp. 7864–5.
52 Zhu Fenghan, 1983; and 1990, pp. 94–104.

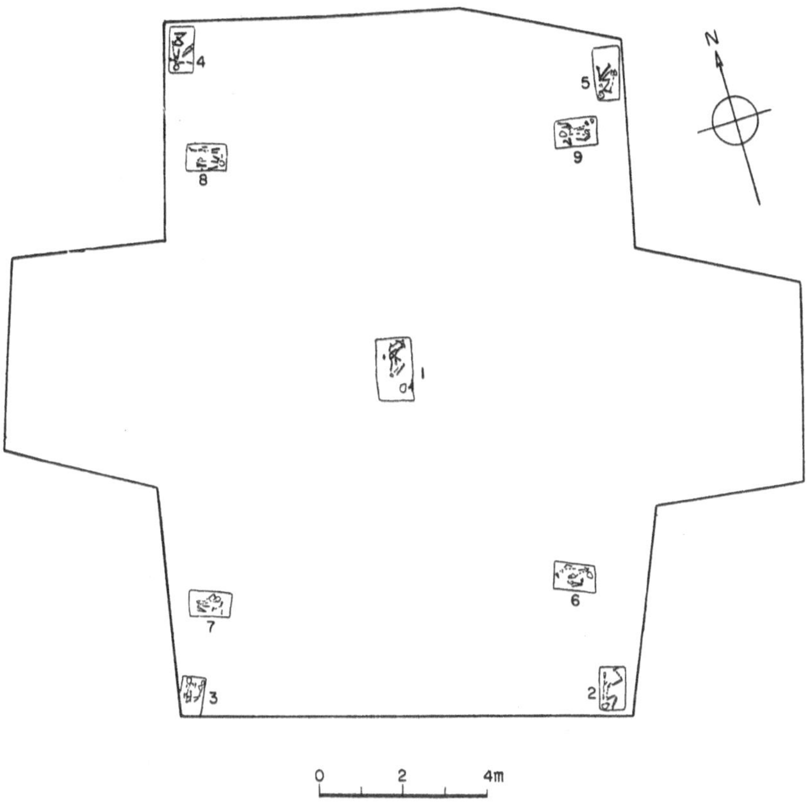

Figure 2.3. Sacrificial burial pits beneath the tomb floor in Shang royal tomb no. 1001 (Liang Siyong and Gao Quxun, *Houjiazhuang*, Taipei: Institute of History and Philology, 1962, vol. 1, Fig. 10, p. 29).

jia 日甲 and *ri xin* 日辛 – a combination indicating association among Shang royal clans.[53] So far, this interpretation of *ya*-shaped clan-signs as an indication of blood ties among the Shang clans from a common ancestry has seemed the most convincing, because the centrality of the royal lineage – the "lineage of us" – seen in the inscriptions is consistent with the textual and archaeological evidence.

These three groups of materials together reveal that the royal ancestral line was the center of the cosmos, and that this center had to be reproduced repeatedly through material construction and ritual action.

53 Barnard, 1986, pp. 141–52.

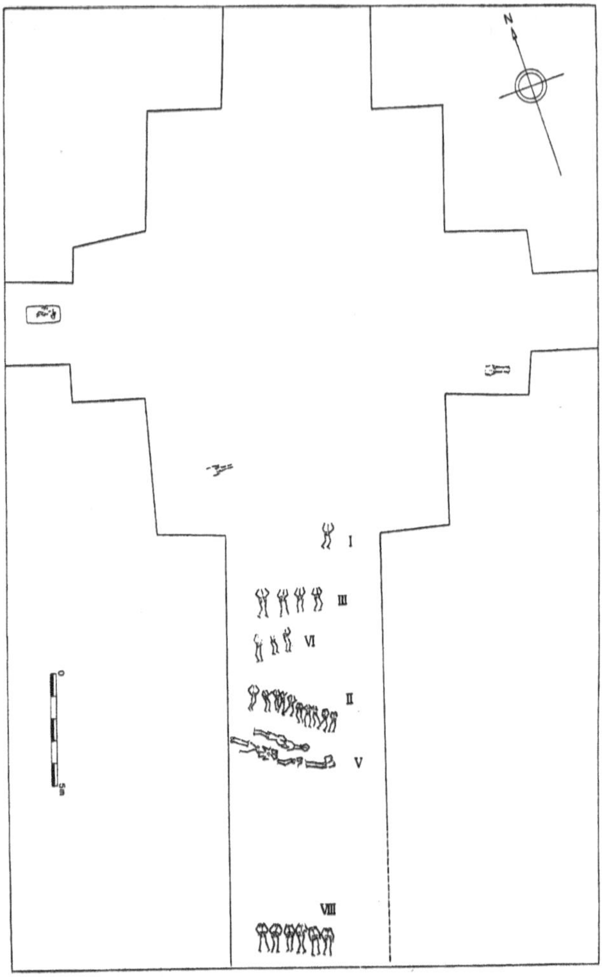

Figure 2.4. Sacrificial human bodies without heads buried in the pit and the western and southern ramps of Shang royal tomb no. 1001 (Liang Siyong and Gao Quxun, *Houjiazhuang*, Taipei: Institute of History and Philology, 1962, vol. 1, Fig. 14, p. 38).

Both temples and tombs served as sacred spaces for ancestors, the "twin centers of ancestral worship."[54] At the temple was housed the altar of the ancestors of the entire lineage and the means to communicate with them, the bronze vessels. There too the sacrificial rituals were per-

54 Wu Hung, 1995, p. 111.

Figure 2.5. Bronze inscriptions with the "*ya*" symbol (Luo Zhenyu, *Sandai jijin wencun*, Pojuezhai, 1936).

formed. The newly deceased ancestor was located at the center of the tomb, and in the four sides of the chamber and in ramps leading to the four directions the human sacrificial rites were conducted. At the center of *ya*-shaped inscriptions was engraved the symbol of the lineage of "us," the centrality of which was signified by the indication of the four directions around the center of the inscription. Thus we see that what is common to these three groups of materials is not only the *Sifang*-center structure itself, but more essentially their function as defining, symbolizing, and reproducing the centrality of the royal ancestral line in ritual action.

The Cosmology of Ritual Political Actions

Space and Time

Sifang and the center form a structural unity, a three-dimensional cosmology. On the one hand, as we have demonstrated, the center was defined in every dimension by the four *fang*: the political center of "us" was defined by the "others"; the geographical center of the ancestral capital was defined by the four lands surrounding it; and the ritual-cosmological center of the royal ancestral line was defined by the lineage of the others. On the other hand, however, *Sifang* depended on this center to become a layered cosmology. In this cosmology, the center was the royal ancestral line through which the king monopolized communication between Heaven and Earth, with his ancestors as the channel, his court shamans as assistants, and sacrificial rites as his journey. In ritual prayers of both the Shang and Zhou periods, the parallel concepts of "high" and "low" (*shang xia* 上下) were frequently used, with "high" as an abbreviation for the high god and "low" for the lower gods – the royal ancestors.[55] This high-low connection indicates the meeting point of Heaven and Earth in this layered universe.

Political and spiritual superiority was achieved by monopolizing access to this layered *Sifang*-center cosmos. The Heaven-Earth connection had constantly to be establised through rituals. And the center – the access to divine ancestral wisdom – had to be reproduced regularly through ritual action, and reborn physically into the succession of the kings. It was such ritual and political action, therefore, that bound *Sifang* and the center into a cosmology and continued to reproduce it.

Sifang was first of all a spatial structure for ritual and political action. As demonstrated in the previous section, *Sifang* not only classified forces

55 Hu Houxuan, 1959.

and spirits in the universe into a structure of four cardinal directions, it also constituted a spatial structure for Shang ritual communication with the divine world, with the four cardinal directions defining the centrality of the Shang and its ancestors.

Besides providing the spatial structure for ritual, *Sifang* was also indispensable for the structuring of time, the ordering of ritual and political action in a temporal sequence. Most Shang terms for units of time were derived from the names of the rites that were repeated at various cycles of time. The temporal order of the sacrificial rites has been extensively studied since the 1940s. Dong Zhuobin discovered that the most important rituals for ancestor worship were the "five rites." The five rites were conducted regularly in an annual cycle in which one after another of the fifty or so ancestors would receive five kinds of sacrifices in a highly complex order.[56] In the past half century, Chen Mengjia, Shima Kunio, and Xu Jinxong have further explored this topic and modified Dong's reconstruction of the "five rites" system.[57] Based on the most comprehensive catalogue of oracle bone inscriptions – *Jiaguwen heji*, which became available in the 1980s, Chang Yuzhi has critiqued and modified earlier conclusions.[58] What this scholarship reveals, for our study of cosmology, is the crucial link between the concept of time and Shang rituals – the temporal structure ordered the rituals, and rituals actualized the conception of time. This close linkage is manifested in the temporal ordering of Shang rituals which can be summarized as follows:[59]

1. The time within a day was divided into six to eight units, among which the middle of the day, called *zhong ri* 中日 ("when the sun is at the center"), was the time at which the most important ritual and political actions, *da shi* 大事, were undertaken.
2. Ten days formed a cycle of *xun* 旬 or "week," and were signaled by the ten day-signs (stems, *gan* 干). Actually, the ancestors were classified and named after the ten day-signs. For example, the founding ancestor Shang Jia was given *jia* 甲, the first day of the *xun*, as part of his temple name. Not only the deceased kings, but also the deceased consorts of the kings were classified by the ten day-signs. Each day during a ten-day week, a sacrifice was offered to ancestors whose temple name matched the sign of that day – on the first day

56 Dong Zuobin, 1945.
57 Chen Mengjia, 1956; Shima Kunio, 1958; and Xu Jinxiong, 1968.
58 Chang Yuzhi, 1987.
59 This summary incorporates the scholarship on the "five rites" just mentioned. In light of the disagreements on technical details, I have adopted Chang Yuzhi's conclusion, since her study includes the most complete and up-to-date material. See Chang Yuzhi, 1987.

sacrifices to the ancestors with the name *jia*, the second day *yi* 乙, and so on. The ten day-signs, therefore, composed a system of classification of ancestors and was of crucial importance for ritual, succession of kingship, and the marriage system.[60]

3. The unit of a ten-day week was called *xun* 旬. *Xun* itself was a cycle of frequent *xun*-sacrifices. At the beginning of each *xun*, the Shang king would divine for the sacrifice of this *xun*.
4. The ten day-signs combined with the twelve other signs called "branches" (*zhi*) 支 formed a larger cycle of sixty days. Carved in used oracle bones, the sixty-day cycle was the only firmly established, written calendar. Keightley suggests that this calendar was little used by the peasantry, but rather possessed religious value and conferred religious authority.[61]
5. The new moon indicated the beginning of the cycle of a month, in which the moon rites were performed. Eclipses of the moon were considered portents of ill fortune, for which sacrificial rituals were offered.
6. The concept of a year was of prime importance for rituals of ancestor worship. The most important rituals for the ancestors were the "five rites" – called *yi* 翌, *ji* 祭, *zai* 翻, *xie* 劦, and *yong* 彤 – conducted in the cycle of a year. In this cycle, all thirty-one former kings, starting with the founding king Shang Jia (上甲), received five sacrificial rites in the order of the succession of kingship, accompanied by twenty selected female ancestors.

This structure of time was interdependent with the *Sifang* spatial structure, and knowledge of the four directions was correlated to knowledge of time. To describe precisely the interdepencence of *Sifang* as a spatial structure and a structure of time during the Shang period has been a subject for the history of science. Joseph Needham has pointed out that before the invention of the compass (in its primitive form called *zhi nan zhen* 指南針 – "south pointing handle") in Han times, the earlier Chinese, to determine a north-south axis, used the celestial pole indicated by the pole stars.[62] This might explain why the north-south orientation of all Shang royal tombs deviated about ten degrees eastward from the north (see Fig. 2.6). However, technology might not be the best explanation for this eastward deviation. Keightley relates this eastward deviation to the prevalent Shang orientation to the northeast, a cultural rather than

60 For an in-depth study of the temple names of the Shang ancestors, see Kwang-chih Chang, 1963a; 1973. For an alternative explanation of the temple names, see Keightley, 1991.

61 Keightley, forthcoming.

62 Needham and Wang Ling, 1962, vol. 4, pp. 301–33; 1959, vol. 3, p. 206.

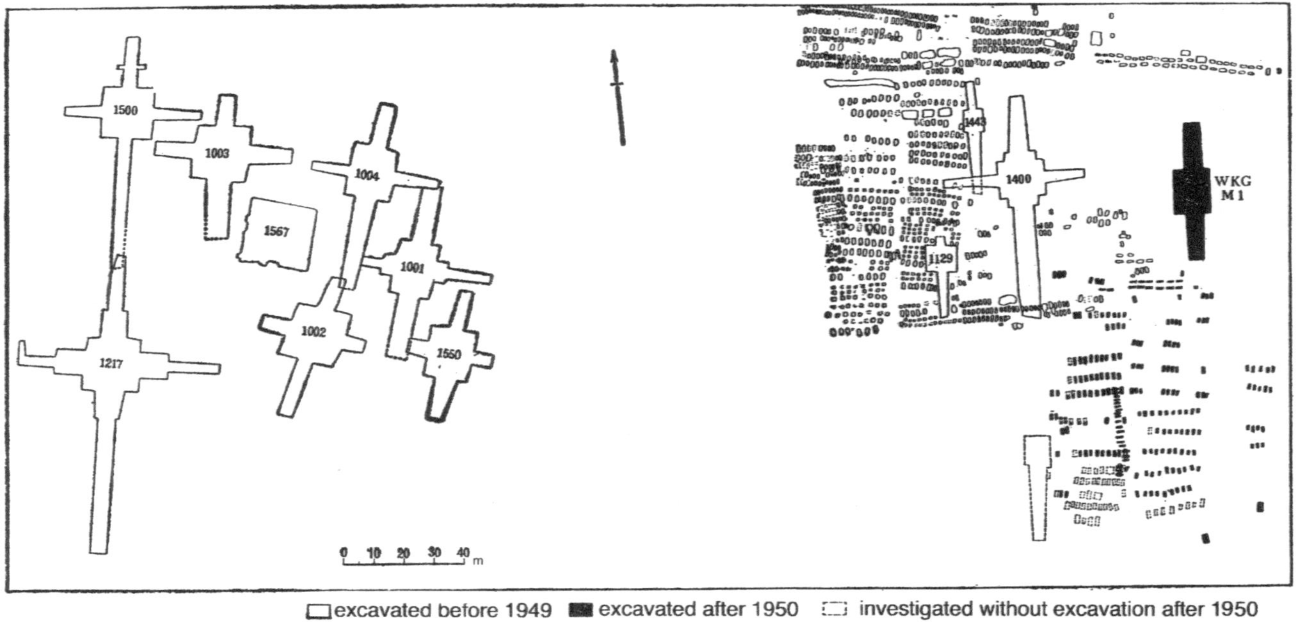

Figure 2.6. Plan of the royal cemetery of the Shang Dynasty at Houjiazhuang, Anyang (Zhongguo kexueyuan kaogu yanjiusuo, ed., *Xin Zhongguo de kaogu faxian he yanjiu*, Kaoguxue zhuankan, series A, no. 17, Beijing: Wenwu chubanshe, 1984, Fig. 61, p. 231).

technical factor that can be explained by the Shang's homeland in the east or the Shang worship of the sun, moon, or stars.[63] Furthermore, the cardinal directions were not necessarily determined by the single method of using pole stars during the Shang period. As Paul Wheatley has suggested, "city walls that were oriented more accurately were probably laid out in conformity with a north-south axis determined by bisection of the angle between the directions of the rising and setting of the sun," rather than by the pole stars.[64] Actually, the rectangular Shang city walls in Yanshi and Zhengzhou (Fig. 2.7), two early Shang palace foundations excavated at Erlitou (Fig. 2.8), and the Western Zhou palace foundations excavated at Fengchu (Fig. 2.9) are in better accordance with true north than the late Shang royal tombs at Houjiazhuang. The slight deviation of the Erlitou foundations was west of true north, as opposed to the easterly deviation when using the pole stars. This suggests that another method of determining cardinal directions, depending on the rising and setting of the sun, may have been used in determining cardinal directions. While no oracle bone records reveal the use of such a technique, Wheatley used later texts – those that appeared roughly between 400 and 200 B.C., including "Kaogongji" 考工記 in *The Rites of the Zhou* (*Zhouli*) 周禮, "Ding zhi fang zhong" 定時方中 in the *Book of Poetry*, and "Yao Dian" 堯典 in the *Book of Documents* – to argue that this method of determining cardinal directions was likely the one used by the Shang and Zhou people in constructing their rectangular cities and palaces.

Further exploring Wheatley's hypothesis, many other scholars have come to the conclusion that the structures of time and space were interdependent in Shang cosmology. The primary pieces of evidence employed in this exploration are inscriptions about *Sifang* and the four winds (*Heji*, 14295 and 36975, quoted earlier in this chapter). Various studies of those inscriptions, including earlier works by Hu Houxuan and Yang Shuda and recent works by Li Xueqin and Feng Shi, agree that *Sifang* and the four winds played a crucial role in the Shang calendar and conception of time. These scholars have demonstrated that the "four winds" of *Sifang* were the gods commanding time, more precisely, the gods of two equinoxes and two solstices.[65] Feng Shi's recent article has incorporated philological analysis with new archaeological evidence from the Neolithic to the early Han period, arguing that *Sifang* and the four winds represent a complete calendrical system as well as a system of standardized time. This system is developed astronomically,

63 Keightley, forthcoming, ch. 6. 64 Wheatley, 1971, p. 426.

65 Hu Houxuan, 1956; Akatsuka Kiyoshi, 1977, pp. 415–443; Ikeda Suetoshi, 1981, pp. 122–137; Li Xueqin, 1985; and Lian Shaoming, 1988.

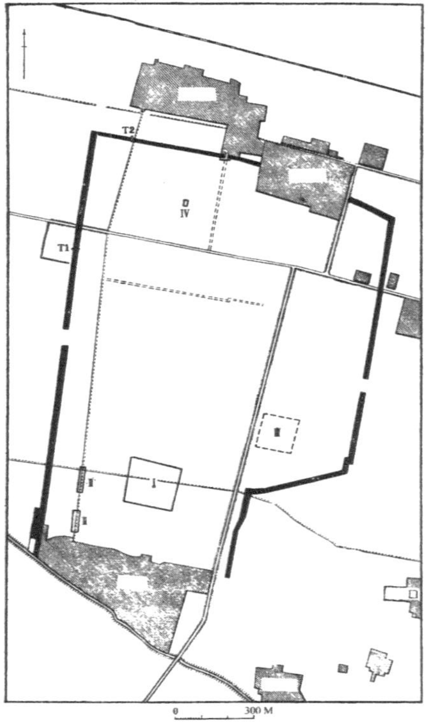

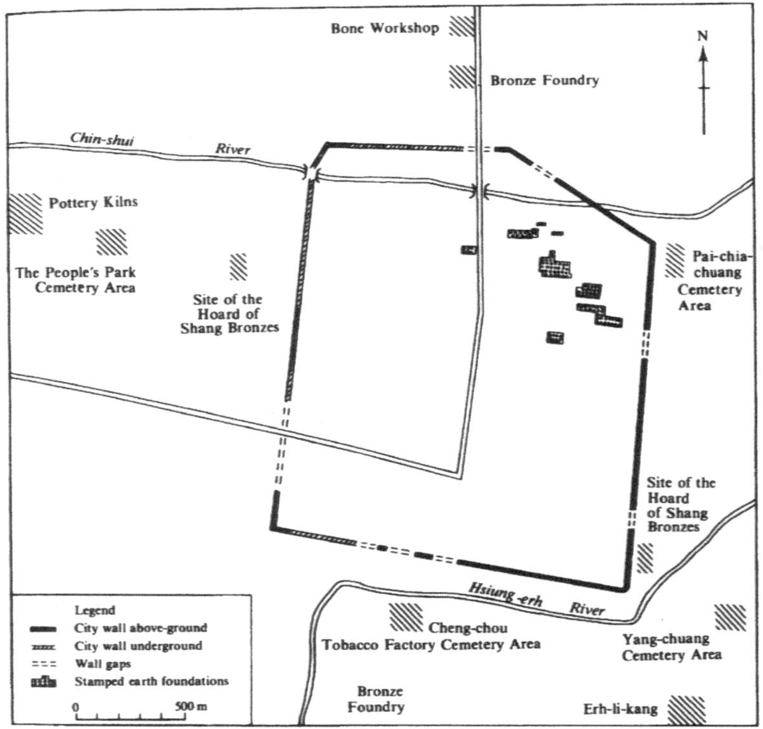

Figure 2.7. Shang city walls (left) in Yanshi (*Kaogu* 6 [1984], p. 490) and (right) in Zhengzhou (An Chin-huai, "The Shang City at Cheng-chou," in *Studies of Shang Archaeology*, ed. Kwang-chih Chang, New Haven: Yale University Press, 1986, Fig. 9, p. 39).

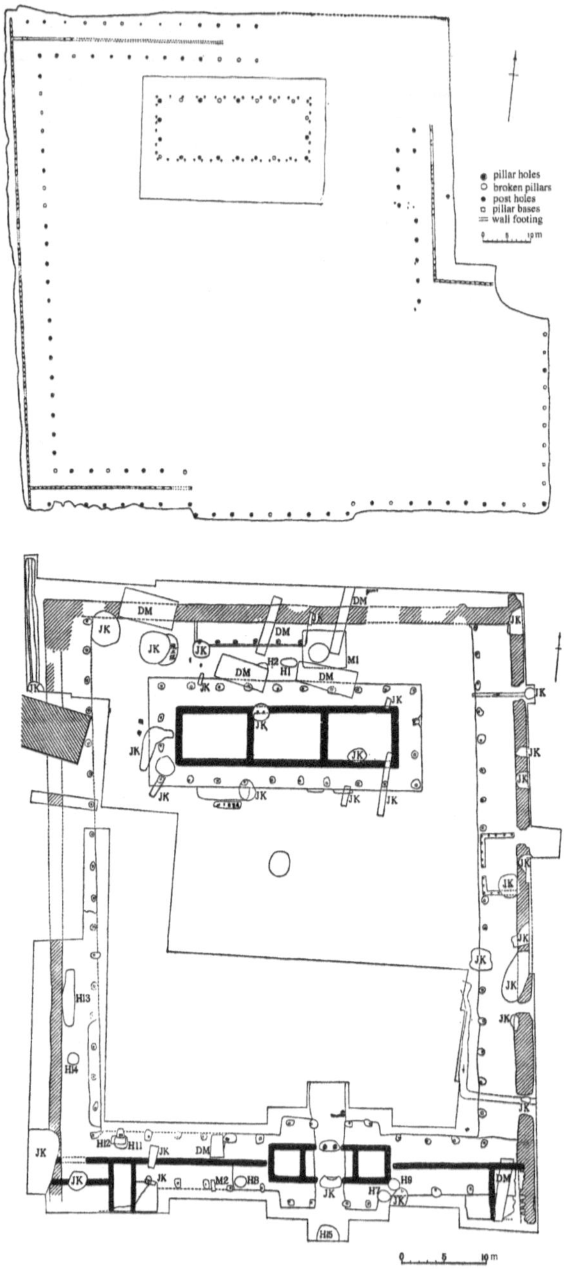

Figure 2.8. Two foundations of palaces excavated at Erlitou, Yanshi, Henan: (top) palace no. 1 (*Kaogu* 4 [1974], p. 235), (bottom) palace no. 2 (*Kaogu* 3 [1983], p. 207).

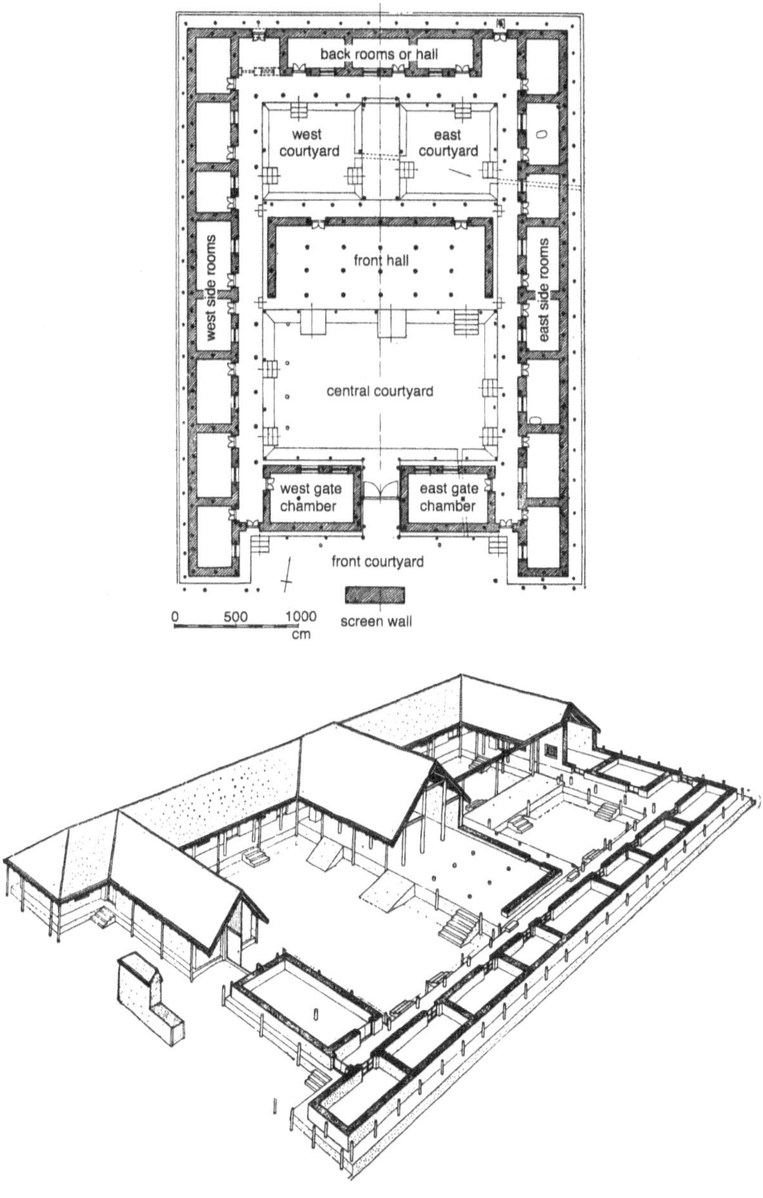

Figure 2.9. A Western Zhou palace compound excavated at Fengchu, Shaanxi province: (top) floor plan and (bottom) reconstruction (Yang Hongxun, "Xizhou qiyi jianzhu yizhi chubu kaocha," *Wenwu* 3 [1981], pp. 24–25).

independent of the conception of the four seasons developed in agriculture. In this system, the four cardinal directions were determined by the positions of the sun at the two equinoxes and two solstices, while the four winds from *Sifang* were the meteorological manifestations of the equinoxes and solstices.[66] Feng's conclusion coincides with that of Chang Zhenguang based on his study of the Shang sun cult. Chang observes that the two sacrificial rites for the sun, "rising sun" and "setting sun" (*chu ri* 出日 and *ru ri* 入日), were always performed on the same day as two inseparable parts of the same ritual. Chang suggests that by measuring the sun's position at sunrise and sunset on an equinox day, the four cardinal directions as well as the four seasons were established.[67]

While the specific Shang techniques for establishing time and space still await proof using more substantial evidence, it is important to remember that this structure of time and space was established to organize ritual activities of ancestor worship, and that in this structure, the concept of center was essential. In Shang inscriptions, the most frequent use of the character for center, *zhong* 中, is in *li zhong* 立中, to establish or erect the center. Based on the graphic of *zhong* 中 symbolizing flags at top and bottom, some scholars believe that this rite involved measuring the sun's shadow by erecting the flag strictly upward when the sun was at the center of the sky, an act of prime importance in regulating time and space, which had to be undertaken by the king personally.[68] Keightley suggests that the phrase could also mean "to set up the standard of the center" or "to set one's self up in the center." He argues that whether it was the king situating himself at the center or establishing some kind of standard there, *li zhong* was a major ritual act of the Shang king, and its importance can be seen from the fact that divination for the performance of this rite was often done twenty or thirty days in advance.[69]

Heaven and Earth

Sifang, with its implied center, is a cosmology of time and space, which symbolizes a unity of Heaven and Earth. The standard symbols of Heaven as a circle and Earth as a square were most often put together to represent a cosmography, in which the circle and the square, or Heaven and Earth, were integrated by the single structure of four cardinal directions and the center. Such cosmography was not simply an

66 Feng Shi, 1994. 67 Chang Zhengguang, 1989.
68 Jiang Liangfu, 1979; and Lian Shaoming, 1996.
69 Keightley, forthcoming, ch. 6, n. 15.

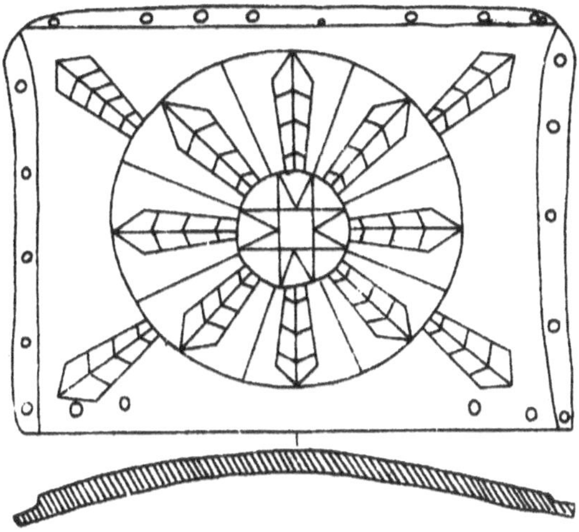

Figure 2.10. Cosmography engraved on a piece of jade unearthed at Hanshan, Anhui (Chen Jiujin and Zhang Jingguo, "Hanshan chutu yupian tuxing shikao," *Wenwu* 4 [1989], p. 15).

image of Heaven and Earth, but rather a structure used to measure the movement of the sun and to regulate time and space.

Sifang as such a cosmology may have had a much earlier existence, as illustrated by the Neolithic cosmography engraved on a piece of jade unearthed at Hanshan, dated around 3000 B.C. (Fig. 2.10). Engraved on this rectangular jade are two circles, one inside the other. The small circle is at the center, with a symbol of the sun inside. Between the small circle and the larger circle there are eight arrows, pointing to the four directions (*Sifang*) and the orientations between the four directions (*Siwei* 四維). The larger circle symbolizes the heavenly movement and the cycle of the seasons according to the sun's changing positions.[70] Both the rectangle, the symbol of Earth, and the circle, the symbol of Heaven, exist in this single cosmography and share the directions and the center. Li Xueqin believes that this diagram is an expression of the ancient Chinese cosmology, composed of a round Heaven and a square Earth with four directions and four orientations, that was repeated by *shi* 式 instruments, sundials, and the Han bronze mirrors called TLV in Western sinology, referring to their cosmological design.[71]

Besides the well-recognized structural continuity between the Shang

70 Chen Jiujin and Zhang Jingguo, 1989. 71 Li Xueqin, 1992–93.

Sifang-center cosmology and the cosmologies of the Han period, a more significant continuity between the two is to be found in their role in constructing sociopolitical reality. The *Sifang*-center cosmology of the Shang functioned as a cultural totality, which not only mystified or legitimated political domination, but also constructed it. Based on the conceptions of time, space, and hierarchy in this cosmology, power was exercised in the rituals of everyday life, constructed in a world of cities, temples, and tombs, and implemented in calendars and geography. This cosmology, therefore, became the embodiment of power relations, with the center occupied by the king – his clan and domains – and the periphery inhabited by the subordinate "others," with time marked through the ritual cycles of royal ancestor worship, and with knowledge of the universe monopolized and reproduced by the ruling clan. In short, it was through this cosmology that power relations and the knowledge of the universe formed a single social composition, and that the domains of the "sacred" and the "profane" fused into a continuity of being in the king's body, an unbroken chain in the royal ancestral line.

Durkheim, based on the comparative study of "primitive religions," defines religion in terms of the separation of the "sacred' and the "profane" – "the sacred thing is par excellence that which the profane should not touch, and cannot touch with impunity."[72] Yet Kwang-chih Chang has challenged such a dichotomy in the case of ancient China, which he describes as a "layered but interlinked world continuum" or a "stratified universe," in which "privileged humans and animals roamed about from one layer to another." Comparing the Chinese vision of nature with that of the West, Tu Wei-ming also points out that the Chinese see the cosmos as continuous with its creator, and see nature as an organic process characterized by continuity, wholeness, and dynamism.[73] The conjugation of power and cosmology became essential to the correlative cosmologies that appeared later. Historians of Europe also have turned to investigating the integration of the divine and terrestrial orders, especially in kingship, which Kantorowicz has characterized as the "king's two bodies."[74]

Yet it was this unity of power and cosmology, the very mode by which power and cosmology combined, that was subject to constant change in historical localities. Based on the evidence from Shang China, it is my argument that the Shang king's body and ancestral line formed the axis for the layered cosmos, an axis that was defined by *Sifang* and that was monopolized by the king for communication with the divine forces.

72 Durkheim, (1915) 1965, pp. 55, 469–70.
73 Kwang-chih Chang, 1986, p. 439; Wei-ming Tu, 1985.
74 For the founding work of this approach, see Kantorowicz, 1957.

Sifang in the Late Bronze Age

Shang cosmology did not directly influence the imperial Chinese culture that was to take shape during the third and fourth centuries B.C., but rather was filtered through the transitional period of the Western Zhou, which replaced Shang's hegemony in 1045 B.C.,[75] until losing it to rising local powers in 771 B.C. During this period of shifting political power, the Zhou inherited, reinforced, and transformed the Shang cosmology of *Sifang*.

The Conquest of Shang

The Zhou were a group living at the periphery of the Shang civilization of the central plain who settled to the west of Shang, in the Wei River valley, around the twelfth century B.C. during the middle of the Shang period.[76] To the Shang, the Zhou were one of the many *fang*, seen in oracle bone inscriptions as "Zhoufang." According to oracle inscriptions of both Shang and Zhou,[77] King Wu Ding of Shang had ordered his lords to attack the Zhou frequently, and the Zhou had at one point become a subordinate polity to the Shang – paying tribute to the Shang, worshiping Shang's founding ancestors, and intermarrying with the Shang.[78] Consequently, Shang oracle bone inscriptions addressed the Zhou sometimes as a *fang* and other times as a lord. During the eleventh century B.C., after conquering many *fang* and allying with many Shang lords, King Wu of Zhou, leading a coalition of lords, attacked the Shang capital and ended the Shang hegemony.

To actually take over a political and ritual center that had lasted nearly a millennium, and to legitimize such a seizure by a former peripheral lord, posed a much greater challenge to the Zhou conquerors than the task of military conquest itself. After winning the battle at the Shang

75 The scholarship on the precise date of the Zhou conquest of Shang is extremely complex and controversial because of the incoherence of the vastly diverse sources of information. This study adopts Shaughnessy's date of 1045 B.C.; see Shaughnessy, 1991, pp. 217–36. For other recent studies on this special subject, see Pankenier, 1981–2, 1983–5, 1995. 76 Shaughnessy, 1999, p. 306.

77 The Zhou oracle inscriptions were discovered at Zhouyuan, Shaanxi, during the 1970s. For the study of these inscriptions, see Wang Yuxin, 1984; Xu Xitai, 1987; Chen Quanfang, 1988; and Hsu and Linduff, 1988, pp. 46–9.

78 The Zhou oracle inscriptions discovered at Zhouyuan recorded that the pre-conquest Zhou sacrificed to Shang ancestors. Zhouyuan, pit number 11, piece numbers 1, 82, 84, 112. For the transcription of these pieces of inscriptions, see Xu Xitai, 1987, p. 175; and Chen Quanfang, 1988, pp. 150–1. For the relationship between Shang and Zhou, see Zhong Bosheng, 1978, p. 20; Kwang-chih Chang, 1980b; Xu Xitai, 1987, pp. 129–37; and Hsu and Linduff, 1988, pp. 41–54.

suburb Muye, and after the death of the last Shang king, the Zhou had to make a shift from the periphery to the center – demographically, politically, and ritually, as well as cosmologically. In making such a shift, the Zhou conquerors did not abandon the Shang tradition. Instead, they continued most of the Shang's religious ideas and ritual practices – including the worship of ancestors, the notion of the high god, and the concept of *Sifang* – as the basis for their political legitimacy.[79]

To claim religious centrality, the Zhou immediately borrowed the Shang concept of a high god, *Di*, to address the newly conquered Shang subjects, and combined *Di* with their own concept of the supreme deity of Heaven, *Tian* 天.[80] The Zhou blamed the last Shang king for having been disrespectful to the gods and spirits and for ignoring the sacrifices, following the advice of women, rejecting men from his own clan, using criminals from *Sifang*, and indulging in drinking alcohol. The Zhou justified their military action as correcting the Shang's betrayal of its own

79 The Shang Zhou transition has been an issue for scholarly debate. Based on transmitted texts, especially Confucian texts, some scholars believe that the Western Zhou marked the fundamental break with the past, and thus should be considered the fountainhead of the Chinese civilization that has continued to the present time. This view is represented by Wang Guowei's monumental work of the 1920s as well as by recent work of Hsu Cho-yun and Katheryn Linduff. See Wang Guowei, (1921) 1984, vol. 10; and Hsu and Linduff, 1988, p. xvii. Yet based on archaeological evidence and oracle and bronze inscriptions, other scholars, as represented by Kwang-chih Chang (1984b), assert that the Zhou continued most of the Shang political and cultural traditions and that therefore the Shang and Zhou constitute a continuous Bronze Age of China, rather than a sharp break.

80 Many scholars agree that the Zhou borrowed the Shang concept of *Di* and used it together with *Tian* to refer to the high god. This agreement is based on rich evidence from both transmitted texts, such as the *Book of Poetry* and the *Book of Documents*, and bronze inscriptions. Chen Mengjia terms the Western Zhou concept of high god "Tian-Di" and points out its continuity and change vis-à-vis the concept of *Di* of the Shang (Chen Mengjia, 1956, pp. 580–2). Herrlee Creel counts 68 uses of *Di* in transmitted texts dated from the Western Zhou, as compared to 228 mentions of *Tian* in the same chapters (Creel, 1970, pp. 494–5). Ikeda believes that *Di* and *Tian* had the same phonetic origin and original meaning, and that they were used interchangeably to represent a deity that incorporated both ancestors and natural powers. (Ikeda, 1981, pp. 25–38). Hayashi Minio also comments on the coexistence of the concepts of *Di* and *Tian* in Western Zhou sources, but believes that they represented different deities (Hayashi Minio, 1989, pp. 9–13). Hsu Cho-yun believes that for the Zhou, "the merging of *Di* and *Tian* was a logical development that created alliance among the groups" (Hsu and Linduff, 1988, p. 108). Yet Shaughnessy cautions that the Western Zhou bronze inscriptions have just two uses of the character *Di*, and that one of them, from the Shi Qiang *pan*, was cast by a family of Shang origin. Numerous uses of *Di* in the *Book of Documents* also appear in the chapters ostensibly addressed to the Shang people. Therefore, how much the Zhou inherited the concept of *Di* is still questionable (correspondence between Shaughnessy and the author, dated December 19, 1997).

tradition.[81] There has been an increasing consensus among modern scholars that the late Shang did deviate from its earlier religious norms – including reducing the high god *Di* to the level of Shang ancestors,[82] replacing divining about uncertainty with highly formulated rituals,[83] abolishing sacrifices to the high god and natural powers altogether, and assuming the title *Di* for the kings.[84] By combining *Di* with their own deity *Tian*, the Zhou not only portrayed themselves as the true heirs of the Shang tradition – thus winning the support of former Shang subjects – but also restored the absolute supremacy of the high god over the Shang ancestors.

Claiming the absolute supremacy of an abstract, universal Heaven, which absorbed the Shang concept of the high god *Di*, provided the foundation for the Zhou's legitimization for replacing Shang – the moral and political concept of Heaven's Mandate. The Zhou claimed that Heaven had shifted its Mandate away from the Shang and given it to the Zhou king, the Son of Heaven, and that the Zhou's conquering of the Shang only served to realize this intention of Heaven. Heaven's inten-

81 The accusation of the last Shang kings was repeatedly stated in Western Zhou bronze inscriptions and transmitted texts, including the *Book of Poetry*, the *Book of Documents*, *Yi Zhoushu*, *Zuozhuan*, and *Shiji*. According to the "Mu shi" chapter in the *Book of Documents*, Tu Cheng-sheng argues that the accusation of Shang by the conquering Zhou king, Wu, was only fourfold, addressed to the Shang lords; it was after the Shang fall that its crime was exaggerated and elaborated upon. See Tu Cheng-sheng, 1992, pp. 319–22.

82 Ikeda, 1981, pp. 47–63. 83 Keightley, 1984, p. 14–18; and 1988.

84 That the late Shang kings deviated from the early Shang norm and changed ritual tradition has long been observed by scholars. As early as the 1940s, Dong Zuobin explained this change in ritual tradition by proposing that there were two schools regarding rituals. The "old school" divined about many subjects, and sacrificed to a broad range of deities, including *Di*, natural forces, *Sifang*, and numerous other deities. The "new school," by contrast, diminished the scale of sacrifice, limited the offerings to only recent ancestors of the Shang, excluded numerous natural deities and even the remote ancestors of Shang, and reduced divination to a mere routine of announcing a well-formulated ritual pattern (Dong, 1944, vol. 1, pp. 2–4). This suggestion has been greatly refined and modified by later scholarship. Kwang-chih Chang verified that Shang ritual manifested a dualism in the archaeological evidence (Chang, 1965–7). Ito Michiharu saw this change in religion and ritual as a sign of Shang losing its ability to consolidate the diverse groups toward the center (Ito, 1975, pp. 58–64, 112–7). David Keightley refined this arguemnt with his study of late Shang divination, noting that late Shang divination became a routine observance of ancestor worship, "forming a routine background of invocation to the daily life of the last two Shang kings, who were now talking, perhaps more to themselves than to the ultra-human powers" (Keightley, 1988, p. 382). David Pankenier in his study of astronomy further verified that late Shang ritual practice deviated from an ancient tradition, and that the Zhou religious and ritual practice after conquest should be seen as correcting late Shang deviation and thus as Zhou continuity of the early Shang traditon (Pankenier, 1995).

tion was not obscure, but had its clear manifestation in the celestial and natural world, as demonstrated by the gathering of five planets in the mansion *Yugui* (Cancer) on May 28 of 1059 B.C.,[85] and the uprooting of a giant tree.[86]

While the Zhou claimed the supremacy of Heaven and created the moral logic of the Mandate of Heaven, they explained the shift of the Mandate in terms of the ruler's *de* 德 – political, symbolic, and moral "power" or "potency" accumulated over time by the ruler.[87] By attributing the shift of the Mandate of Heaven to the living King's "potency" – what he had accumulated through his own action – rather than to his affiliation with a particular ancestral line, the Zhou initiated the long process of creating a concept of universal sovereignty that was not based on a particular ancestral line. This moral logic of the Mandate of Heaven particularly undermined the Shang king's monopoly on access to the divine world through his ancestors, since the living king could lose his potency because of his own bad behavior and thus lose the Mandate to another lineage.

Although the Zhou undermined the authority of the Shang ancestors with the concept of the Mandate of Heaven, they did not deny the religious and cosmological centrality of ancestor worship. On the contrary, the Zhou kings based their political domination on the cosmological centrality of their own royal lineage. The Western Zhou bronze inscriptions provide abundant evidence of the central position of ancestor worship.[88] Immediately after the military conquest, as the inscription on the Dafeng *gui* vessel describes, King Wu offered a series of ritual sacrifices – to the *Sifang*, to the high god *Di*, and to the ancestors. Most interestingly, the king offered the grand ancestor worship rites of the Shang people called *Yi* (衣 or *Yin* 殷) sacrifices to the Zhou ancestor King Wen, as well as to the Shang ancestors:[89]

85 David Pankenier has done extensive analyses of ancient astronomy and the relation between astronomical phenomena and Shang and Zhou legitimation of power (Pankenier, 1995). Pankenier also suggests that, rather than being a Zhou innovation, the concept of Heaven's Mandate – the correlation between celestial or natural signs and the changing of dynasties – existed long before Zhou conquered Shang (Pankenier, 1995, p. 172).

86 Hsu and Linduff, 1988, p. 104.

87 Constance Cook recently summarized the scholarship on *de*, saying that *de* in the Western Zhou context indicated the "power" or "potency" that accumulated over time with increased rewards and prestige. See Cook, 1995, pp. 246–7.

88 In fact, the nature of bronze inscription itself proves the importance of ancestor worship in Western Zhou. As Lothar von Falkenhausen has pointed out, the inscriptions are essentially religious documents intended to communicate with the ancestral spirits. See Falkenhausen, 1993, pp. 146–52.

89 The translations of the bronze inscriptions in this chapter are my own, unless otherwise specified. Dafeng *gui* is also called "Tianwang *gui*." The transcription of the inscrip-

乙亥，王又大豐
王凡三方
王祀于天室，
天亡又王
衣（殷）祀于王不顯考文王
事喜上帝
……
不克三衣王祀
……

On *yihai* day, the king performed grand ceremonies.
The king sailed to the three *fang.*
The king sacrificed at the Heavenly Chamber, descended.
Tian Wang assisted the king,
Offering *Yi* [*Yin*] sacrifices to the illustrious deceased father, King Wen.
Sacrifices to the God on High
……
Three times sacrifices to the kings of Yin [Shang].
……

This inscription reveals that King Wu's grand ritual sacrifices were made to three major objects: *Sifang*, the high god *Di*, and ancestors. The sentence "The king sailed to the three *fang* 王凡三方" has inspired many different interpretations. Lin Yun interprets the "three *fang*" as the four *fang* (*Sifang*) missing a stroke.[90] Ma Chengyuan accepts "three *fang*" as intended.[91] Hsu adopts the earlier suggestion by Ito Michiharu and Chen Mengjia, interpreting the sentence as evidence of the Zhou seeing themselves as the western *fang*, making the offering to the other three *fang*.[92] Shirakawa and Ma Chengyuan interpret the sentence as describing the king performing a specific ritual, *Dafeng* 大豐, in the boat, sailing to the three directions in the Great Pool (*Dachi* 大池).[93] Whether "王凡三方" represents ritual offerings or a specific ritual performed in a boat in the Great Pool, it involves paying ritual respect to three *fang* or directions, and the interpretation that the Zhou king sees himself representing one of the four *fang*, the West, is still feasible. As I will demonstrate later, the Zhou at the time of conquest identified themselves as "men of the West" and called their lord the "lord of the Western *fang* (*Xifang Bo* 西方伯)." The last sentence quoted, "三衣王祀," has been a

tion is adopted from Ma Chengyuan, 1988, vol. 3, no. 23, p. 15; and Shirakawa, vol. 1, no. 1, pp. 1–38.

90 Lin Yun, 1993. 91 Ma Chengyuan, 1988, vol. 3, no. 23, pp. 14–15.

92 Hsu and Linduff, 1988, p. 99; Ito Michiharu, 1978, pp. 48–9; and Chen Mengjia, 1955a, pp. 14–15, 152.

93 Shirakawa, vol. 1, entry no. 1, pp. 1, 9–14; and Ma Chengyuan, 1988, vol. 3, no. 23, pp. 14–15.

puzzle, and no conclusive interpretation has been reached.[94] Both Ma Chengyuan and Hsu Cho-yun interpret it as a grand rite of ancestral worship of the Shang (*Yi* 衣 as a phonetic borrowing of Yin 殷). But Ma Chengyuan believes that the Zhou performed this Shang ritual to the Zhou ancestor, King Wen, and Hsu believes that the rite was offered to the Shang ancestors collectively at the Shang ancestral temple.[95] Since the preconquest Zhou did sacrifice to the Shang ancestors,[96] the worship of Shang ancestors may have continued under the Zhou conquerors as a means of winning the support of the Shang lords.[97]

While the Zhou conquerors preserved the centrality of ancestor worship, they gradually replaced the authority of the Shang royal ancestors with that of their own ancestors, and substituted Zhou's own myth of its divine origin for that of the Shang. In the myth of the Zhou ancestry and genealogy found in late Zhou inscriptions and transmitted texts, the Zhou ancestors are described, very much like those of the Shang, as ascending and descending between the divine and human worlds, living in the world of *Di* as *Di*'s guests. As the song for King Wen – the father of King Wu and believed receiver of the Mandate – states:[98]

King Wen is on High.
Oh bright is he in Heaven [*Tian*].
......
King Wen ascends and descends,
On the left and right of the high god [*Di*].

To communicate with their own ancestors who lived in the divine world beside the high god, the Zhou continued most of the Shang's elaborate rituals, including the *Yi* sacrifices mentioned above.[99] Shaughnessy has recently argued that the Zhou also continued to use Shang day-signs, the ten stems, to classify their own ancestors, and that it was not until

94 Chen Mengjia and Ma Chengyuan interpret this sentence as "terminate the sacrifices to the Yin kings," considering the word "three" 三 as a mis-script of *qi* . . . 气. (Chen Mengjia, 1955, p. 153; Ma Chengyuan, 1988, vol. 3, no. 23, p. 14.) For other interpretations, see Shirakawa, vol. 1, no. 1, pp. 25–8.

95 See Ma Chengyuan, ibid., and Hsu and Linduff, 1988, pp. 99–100.

96 Zhouyuan, pit number 11, piece numbers 1, 82, 84, 112. For the transcription of these pieces of inscriptions, see Xu Xitai, 1987, p. 175; and Chen Quanfang, 1988, pp. 150–1.

97 Chen Quanfang states that the Zhouyuan inscriptions on piece number 1 in pit number 11 describe King Wu's sacrifices to Shang ancestors with two consorts of the last Shang king, and that this piece was likely inscribed thirty days after King Wu conquered Shang and returned to Zhouyuan (ibid.).

98 For the full translation, see Legge, 1963, vol. 4, pp. 428–31.

99 Chen Mengjia listed eighteen kinds of rituals that both the Shang and the Zhou practiced, *Yi* sacrifice being one of them; he concluded that ninety percent of the Zhou ritual names derived from the Shang. See Chen Mengjia, 1936.

the middle Western Zhou that they invented their own institution of *shifa* 謚法 titles for the deceased ancestors.[100]

Shifting to the Center

The Zhou thus inherited the twin pillars of the Shang religion – the concept of a high god (*Di/Tian*) and ancestor worship – as the ideological foundation of their political power. But these twin pillars of Shang were built on the *Sifang* cosmology. The Dafeng *gui* inscription quoted earlier reveals that, after the conquest, one of the Zhou's objects for ritual offering or respect was *Sifang*, together with the high god and ancestors. The Mandate of Heaven meant not only the right to rule, but also the occupation of the center that was defined by *Sifang*, the center of the earth where the Shang king formerly resided, the center of the divine world where royal ancestors ascended and descended, and the center of the multilayered cosmos where the living king communicated with the divine world through rituals. Without possessing such a political, religious, and cosmological center, the Zhou did not truly possess the Mandate of Heaven.

At the time of the conquest the Zhou had an identity as "men of the West." Before the conquest, the Zhou addressed their own lord as the "the lord of the Western *fang* (*Xifang Bo* 西方伯)," according to oracle inscriptions from Zhouyuan.[101] After the conquest, as Tu Cheng-sheng points out, the Zhou believed that the Mandate was shifting westward onto the Zhou.[102] Many chapters in the *Book of Documents* reflect the Zhou's self-identity as inhabitants of the West, and the idea of the Mandate shifting to the West. "Tai shi 泰誓" (The great declaration), a chapter dated after the Western Zhou, recounts that King Wu called out to his people as "my valiant men of the West" to realize Heaven's course of duty to punish the Shang and claim the Mandate. He explained that the Mandate has descended to the Zhou because his deceased father, King Wen, illuminated "the Four Quarters (*Sifang*), and shone signally in the Western region."[103] King Wen's illumination made the god's head turn to the West. The song "Huang yi 皇矣" (Great is God) in the *Book of Poetry* seems to continue King Wu's line, saying that in seeking someone to replace Shang to hold his Mandate, the God on High "sought and considered . . . hating all the great [states] (*si guo* 四國), He

100 Shaughnessy, 1997.
101 Oracle bone inscriptions from Zhouyuan, pit number 11, piece number 82. For the transcription of these pieces of inscriptions, see Xu Xitai, 1987, p. 175; and Chen Quanfang, 1988, pp. 150–1.
102 Tu Cheng-sheng, 1992, pp. 322–30.
103 Legge, 1963, vol. 3, pp. 294–7.

turned His kind regard on the West."[104] This kind of rhetoric reflects the fact that, after the conquest, the Zhou still saw the Shang as the central and ultimate point of reference for the Zhou's existence, referring to their own homeland as "my western land" and referring to themselves as "the people of the western land." Examples of such self-identity are found in many early chapters in the *Book of Documents*, including "Tai gao 泰誥," "Mu shi 牧誓," "Kang gao 康誥," "Da gao 大誥," "Jiu gao 酒誥," and so on.[105]

To shift from the lord of the West to the holder of Heaven's Mandate at the center was an endeavor that involved demographic movement, political organization, military campaigns, and ritual reforms, all of which took generations of Zhou kings to fulfill. Demographically, the Zhou conquerors immediately moved major Shang lords and loyalties to the West, in groups of entire lineages. This move was meant to reduce the power of the Shang groups of the central Yellow River valley, and to absorb the professionals of the Shang into the Zhou regime.[106] "Duo shi 多士" in the *Book of Documents*, an edict of the Zhou King to the Shang immigrants, clearly states the justification for this migration: "I declare to you, the numerous officers . . . that I have removed and settled you in the West. It was not that I, the one man, considered it a part of my virtue to make you untranquil. The thing was from the decree of Heaven."[107] While it is not clearly stated in this passage where exactly the "West" is located, archaeological findings show evidence of Shang groups migrating to Zhouyuan, the homeland of the Zhou. These findings include large sets of bronze vessels excavated at Fufeng 扶風 county of Shaanxi in 1975, 103 pieces altogether, among which 74 are inscribed. These vessels belonged to a family named Wei for several generations. The most important piece is the Shi Qiang *pan*, which bears a long inscription tracing the ancestry of the Wei family back about five generations to the Shang state of Wei. An ancestor of the Wei family served as a scribe in the Wei state. At the time of the Zhou conquest, he came to the court of King Wu and then migrated to the Zhou homeland in Zhouyuan. His descendants served the Zhou as scribes.[108]

104 Legge, 1963, vol. 4, pp. 448–9.
105 See Legge, 1963, vol 3, pp. 294, 297, 300–4, 365, 399.
106 For an extensive analysis of the relocation of the Shang lineages and the position of these Shang lineages in Western Zhou, see Hsu and Linduff, 1988, pp. 113–23; Tu Cheng-sheng, 1992, pp. 352–94, 509–42.
107 Legge, 1963, vol. 3, pp. 459–60.
108 This archaeological evidence and the bronze inscriptions are extensively quoted and interpreted or translated in several recent studies of the Western Zhou; see Hsu and Linduff, 1988, pp. 113–23; Shaughnessy, 1991, pp. 1, 185–92; and Tu Cheng-sheng, 1992, pp. 352–94, 509–42.

While the Zhou ruling clan moved Shang groups westward, they planned to occupy the central region themselves. Hsu Cho-yun argues that King Wu at the time of the conquest was already preoccupied with moving into the central region, using inscriptions on a bronze vessel, the He *zun,* as the key piece of evidence.[109] The He *zun* was unearthed at Baoji, the interior of Zhou's homeland. It was inscribed with the record that King Wu, after conquering the great capital of Shang, reported to Heaven that he planned to reside in the "central region, 中或."[110] This plan involved constructing a new capital, Luo 洛 or Chengzhou 成周, at the junction of the Luo and Yellow Rivers west of the old Shang capital. The plan was executed after King Wu's death by Dukes Shao and Zhou, two brothers of King Wu and the most powerful leaders of the Zhou, and was fulfilled during King Cheng's reign.[111]

The construction of the new capital in the central region was strategically significant for military and political control of the central plain and the remaining old Shang lineage groups, as many historians have pointed out,[112] and it was also of cosmological, religious, and symbolic importance for the Zhou's dominance of *Sifang* or "all under Heaven 天下." It was from this new capital in the central region that the Zhou kings fully received the Mandate of Heaven in commanding the *Sifang.* Several chapters in the *Book of Documents* recount the construction of the new capital, all stressing such cosmological and symbolic significance. "Kang gao 康誥" (The announcement to the prince of Kang) states that the campaign to build the new capital was itself an act that absorbed the people from *Sifang* toward the center:[113]

> Duke Zhou commenced the foundations and proceeded to build the new great city at Luo of the eastern states. The people from the Four Quarters (*Sifang*) assembled in great harmony.

109 Hsu and Linduff, 1988, pp. 96–101.

110 He *zun,* Shirakawa, vol. 48, entry no. 1, p. 171; Ma Chengyuan, 1988, vol. 3, vessel no. 32, pp. 20–21. Ma Chengyuan, Hsu Cho-yun, and Tu Cheng-sheng, among others, interpret the characters 中或 as the "central kingdom" 中國, the concept of China as the center of the civilized world. See Hsu and Linduff, 1988, p. 99; and Tu Cheng-sheng, 1992, p. 452. But Zhao Boxiong argues that the characters 中或 in the inscriptions most often referred to a region 域, and that the "central region" 中或 was used in reference to the four regions – Northern, Southern, Western, and Eastern. (See Zhao Boxiong, 1990, pp. 160–71.) Zhao's argument confirms my findings regarding *Sifang, si guo,* and *si tu* in Western Zhou inscriptions, that 中或 is a geographic concept defined by the four directions, and that the concept of "central kingdom" embodies too many meanings of later times that were not yet present in the Western Zhou. Therefore, I translate 中或 as the "central region" in order to avoid imposing later meanings on the Western Zhou inscriptions.

111 Hsu and Linduff, 1988, pp. 123–26.

112 Hsu and Linduff, ibid.; and Tu Cheng-sheng, 1992, pp. 352–94, 509–42.

113 Legge, 1963, vol. 3, p. 381.

"Luo gao 洛誥" (The announcement concerning Luo) further elaborates how the construction of the new capital in the central region was crucial for the Zhou's maintaining the Mandate of Heaven. It records statements by Duke Zhou and King Cheng concerning the construction of the new capital. To summarize its contents,[114] Duke Zhou divined about building the new capital because he worried that the young King Cheng "appeared as if he could not presume to determine the founding and the fixing of our Mandate of Heaven." Duke Zhou therefore "made a great survey of this eastern region" in order "to find the place where he [King Cheng] might become the intelligent sovereign of the people." After Duke Zhou presented to the young king the results of his divination, indicating the location of Luo, and a map of the region, he instructed King Cheng to initiate the new capital by "first employing the ceremonies of Shang, and sacrificing in the new city."

Duke Zhou considered the occupation of the central region by the Zhou king as essential to the Zhou's maintaining the Mandate and becoming sovereign over all under Heaven. By constructing the new capital, Duke Zhou sought to ensure that the Zhou king would finally occupy the ritual and religious center, taking over the monopoly of communication with the divine world from Shang by performing the ritual sacrifices at the center. While the "ceremony of Shang" would continue, it would be performed by the legitimate Zhou king at the new capital.

King Cheng, responding to Duke Zhou, announced that his role from then on was to "display a brilliant merit like that of King Wen and King Wu. Reverently respond to the Mandate of Heaven, harmonize and long preserve the people of the Four Quarters (*Sifang*), and settle their multitudes here." King Cheng thus declared a threefold definition of kingship – its source from the Zhou ancestors, its authority from the Mandate of Heaven, and its ultimate function of governing the Four Quarters from the center.

In this context, *Sifang* was a political concept referring to "all under Heaven," magnetized towards the center. It was the place where the dominance from the center should extend, where the order set at the center should reach, and where the people should be subjected to the center. In this single chapter, the concept of *Sifang* is used five times, all confirming such a concept of political geography. For example, King Cheng, charging Duke Zhou to remain in support of the king, describes the current situation as: "Order has been initiated throughout the Four Quarters (*Sifang*) of the empire." And Duke Zhou responds: "Good gov-

114 Ibid., pp. 434–52.

ernment here will make you indeed the new commander of the Four Quarters," and "by the government administrated in this central spot, all parts of the empire will be conducted to repose, and this will be the completion of your merit."[115] Presiding over the new capital and ruling over the Four Quarters from the center, according to Duke Zhou, were to be the essence of kingship.

Transforming Sifang

While they inherited the concept of *Sifang*, the Zhou also transformed it, as they did most Shang ideas and practices. As already demonstrated, the Zhou used *Sifang* to refer to a political geography of "all under Heaven," where the Zhou's Mandate prevailed. The Zhou thus transformed the Shang *Sifang* cosmology in two respects. First, the Shang's *Sifang* was a cosmological structure that classified all alien forces, alien polities and their spirits, and the forces of nature. Such a cosmology also functioned as a structure of ritual communication between the Shang king and the divine world. In the Zhou rhetoric, by contrast, *Sifang* seldom appears as a structure classifying the myriad natural forces, or as a subject of divination for unknown forces coming in upon the Zhou, but primarily as a concept of political geography. The usage of *Sifang* in Zhou material almost uniformly refers to "all under Heaven," where the Mandate of Heaven extends. Zhao Boxiong has made a similar argument concerning the concept of *bang* 邦 in Western Zhou. He argues that in Zhou materials, the concepts of many *bang* (*duo bang* 多邦, *wan bang* 萬邦, *shu bang* 庶邦), great and small *bang* (*da xiao bang* 大小邦), and *Sifang* are all expressions of the Zhou concept of the universe – "all under Heaven" – in which the Zhou and the many *bang* coexist.[116] In this universe, as I shall demonstrate, the Zhou king resides at the center and extends his sovereignty over the multiple groups in *Sifang*. Thus, the *Sifang* that was heterogeneous and alien to the center in Shang cosmology became homogeneous and subject to the center in the political ideal of the Zhou.

This conceptual transformation of *Sifang* directly reflects Zhou military and political actions, through which the Zhou actively transformed the people of *Sifang* from alien polities into subjects of the Zhou king, the Son of Heaven. The recurring references to *Sifang* found in the Western Zhou bronze inscriptions illustrate such a transformation of "*Sifang*," both as a concept and as a political reality. An important usage

115 Ibid. 116 Zhao Boxiong, 1990, pp. 13–19.

of *Sifang* in Zhou inscriptions is found in the accounts of the Zhou's military campaigns that brought and kept *Sifang* under Zhou dominance, expressed in such phrases as "campaigning throughout *Sifang* 遹征四方" and "extending to *Sifang* 匍有四方." These accounts reveal that the relationship of the center – Zhou – to *Sifang* was that of the center's continued conquest of the people of *Sifang*.

The best example of military action that converted *Sifang* from aliens to subjects is the inscription on the Shi Qiang *pan* vessel. In recounting the genealogy of the Wei family in relation to the Zhou royal house, the inscription provides a full account of the tale of the Zhou – how the Zhou received the Mandate and possessed the Four Quarters.[117] Starting with King Wen's receiving of the Mandate, it describes how King Wu "proceeded and campaigned throughout the Four Quarters 遹征四方, piercing Shang and governing its people." King Wu not only conquered the Shang, but also defeated the Di groups in the North and therefore was "eternally unfearful of" them, and "attacked the Yi minions to the East." Generations later, King Zhao campaigned in the South: "[He] broadly tamed Chu and Jing; it was to connect the southern route." This inscription describes precisely how *Sifang* was brought under Zhou control through military campaigns. If we consider the West to be the Zhou's own homeland, consolidated before the conquest of Shang, the "campaign throughout the Four Quarters" was composed of three major stages – defeating the Di groups to the North, attacking Yi groups to the East, and taming Chu and Jing to the South. This great military achievement of generations of Zhou kings was frequently glorified in an abbreviated form in bronze inscriptions, such as "King Wu succeeded King Wen in building the state and eradicating its evil, extending to the Four Quarters 匍有四方, and governing its people,"[118] or simply "opened up *Sifang* 闢四方 and glorified the Mandate of Heaven."[119]

As the Zhou campaigned throughout *Sifang*, they extended their political order to those areas they conquered, transforming alien groups into subjects of their rule. Two recent studies of the Western Zhou, by Hsu Cho-yun and Tu Cheng-sheng, extensively discuss this political order in terms of "Zhou feudalism" (*fengjian* 封建). These authors commonly point out that the Zhou extended their rule to a vast area much larger and more diverse than the Shang domains by means of *fengjian* –

117 The transcription is adopted from Shirakawa, vol. 50, entry no. 15, p. 335. The English translation is adapted from Shaughnessy, 1991, pp. 185–6.
118 Da Yu *ding*, Shirakawa, vol. 12, entry no. 61, p. 647.
119 Lubo zhong *gui*, Shirakawa, vol. 17, entry no. 92, p. 209.

a system of delegated authority based on contracts and personal bonds through the Zhou kinship system. By entrusting newly conquered groups to members of the Zhou clan through ritual contracts, the Zhou created a multilayered hierarchy of power that incorporated the Shang nobility, the native populace, and themselves in a tripartite coalition in which a small number of Zhou royal kinsmen ruled over the much larger population of the conquered.[120]

Feudal lords maintained the authority of the Zhou clan over the *Sifang* peoples by their constant military presence as vassals keeping *Sifang* in subjection. This function of the feudal system was reflected in appointment records – bronze inscriptions cast to record the ritual contracts by which the Zhou king endowed aristocrats with the authority to command conquered lands and peoples.[121] One such appointment record is the inscription on the Ban *gui* vessel, which records the Zhou king's command that Duke Mao "set a standard for *Sifang* 乍四方亟." Such a standard involved the duke's "possessing the regions of Fan, Shu, and Chao," leading military attacks upon the eastern regions, and ultimately having "pacified the eastern regions" within three years. Since then, it continues, there have been none "who do not submit to Heaven's majesty."[122]

While submitting the people of *Sifang* to the Zhou's heavenly majesty through the vassal lords' military presence, the Zhou also imposed its bureaucratic and hierarchical order over *Sifang*. Herrlee G. Creel has proposed that the early Western Zhou already had a prototype bureaucracy, which was developed into a complex administrative machine during the middle and late Western Zhou.[123] Inscriptions on the Ling *yi* provide one clue to how such an administrative machine and hierarchical system dispensed the king's orders from the center to *Sifang*. The inscription records that the Zhou king commanded the son of Duke Zhou, named Ming Bao, to "govern the three [chargés d'] affairs and the Four Quarters and to give them administrative officers 尹三事四方，受（授）卿事寮." The "three [chargés d'] affairs" in this

120 Hsu and Linduff, 1988, pp. 147–85; and Tu Cheng-sheng, 1990, pp. 479–508.

121 Edward Shaughnessy analyzes the structure of the Zhou bronze inscriptions as fourfold: (1) date and place notation; (2) event notation; (3) a list of gifts given by the king to the endowed aristocrat; and (4) dedication of the vessels to ancestors. Shaughnessy, 1991, pp. 73–87.

122 For the transcription and interpretation of the Ban *gui*, see Shirakawa, vol. 15, entry no. 79a, p. 34. For English translations, I have adapted from Shaughnessy, 1991, 251–6.

123 Creel, 1970, pp. 117–21. For recent studies of the Western Zhou government, see Zhang Yachu and Liu Yu, 1986; Hsu and Linduff, 1988, pp. 227–57; and Zhao Boxiong, 1990.

context is understood as a general term for administrative officialdom,[124] and the "Four Quarters (*Sifang*)" as referring to all the domains subject to the Zhou.[125] Ming Bao was given the highest administrative position, commanding both the bureaucratic machine of the Zhou and all the vassals in *Sifang*. The later part of the inscription records that Ming Bao, in fulfilling the king's command, traveled to Chengzhou – the capital on the central plain – where he carried out the king's command throughout the entire administrative and hierarchical system. He dispensed the king's order to the personnel of the central administration, including the "many officers 眔卿事寮," the "many governors 眔者尹," "resident heads 眔里君," and "the hundred artisans 眔百工." He also dispensed the order to the personnel of the vassals in *Sifang*, including the "many lords 眔諸侯," "fieldsmen 侯甸," and "sires 男." When the king's command had been successfully carried out by the administration ("three affairs") and communicated to the Four Quarters of the world (*Sifang*), Ming Bao concluded his task with a series of sacrificial rituals at ancestor temples, and then returned from his journey.[126] The Ling *yi* inscription reveals a bureaucratic and hierarchical system through which the Zhou king's orders were administered throughout the Zhou world. The Zhou subjected the people of *Sifang* to its civil orders, which were enforced by military coercion.

The Zhou transformation of *Sifang* – from a cosmology classifying alien forces and spirits to a cosmology depicting Zhou's own domain of governance – illustrates the Zhou's political progression. As Hsu Cho-yun has summarized, "the Zhou royal house developed as the head of a coalition of several constituents, gradually evolved into a suzerainty ruling over feudal vassals, and finally turned into a monarchy with a rather effective governing body."[127] Such a political progression was achieved by constantly transforming the people of *Sifang* from alien "others" to subjects of the Zhou.

This transformation indicates the changing relationship between

124 Guo Moruo believes that "three affairs" is a general term for officialdom. For Guo's interpretation and many others, see Shirakawa, vol. 6, no. 25, pp. 287–89. The three affairs were also called the "three supervisors" or "ministers" – supervisor of works, supervisor of the horse (military), and supervisor of land. See Shaughnessy, 1992, p. 196, no. 7; and Hsu and Linduff, 1988, pp. 237–8.

125 Shirakawa persuasively interprets "*Sifang*" in bronze inscriptions as "all domains and vassals of the Zhou." He also distinguishes the three affairs as a general term from the internal domain of the Zhou, called "*neifu*," and from the Four Quarters as the external domains, called "*waifu*." See Shirakawa, vol. 6, no. 25, pp. 287–95.

126 The analysis of the inscriptions on the Ling *yi* is based on the transcription and interpretation by Shirakawa, vol. 6, no. 25, pp. 276–307; the English translation is adopted from Shaughnessy, 1992, pp. 194–8.

127 Hsu and Linduff, 1988, p. 146.

Sifang and the center. In the Shang vocabulary, *fang* or "many *fang*" referred to the polities that were alien to the Shang, and *Sifang* classified the natural forces and spirits that were outside of the central line of ancestral spirits of the Shang. *Fang* and *Sifang*, therefore, defined Shang political and religious centrality in terms of otherness, outerness, or periphery. In the Zhou rhetoric, by contrast, *Sifang* referred to all the regions and groups subject to – or that should be subject to – the Zhou's Mandate of Heaven. The Zhou thus transformed and reinvented *Sifang* through their political practice.

Continued Centrality

The Zhou transformation of *Sifang*, as a concept and as a political reality, shows strong continuity with the Shang *Sifang* cosmology rather than a rupture.[128] Like the Shang, the Zhou defined the absolute superiority of the center – the king and his ruling clan – by the inferiority and peripheral nature of *Sifang*. The Shang buried the captives from the alien polities of *Sifang* in the four ramps of their ancestral tombs, signifying the absolute superiority of the Shang ruling clan by slaughtering the alien others as sacrificial animals. The Zhou reinforced the centrality of the king and his clan by transforming the concept of *Sifang* from alien forces to the possession of the king, thus defining the centrality of the Zhou clan by the subjection of *Sifang*, rather than by their alienation. The Zhou claimed that their ancestor King Wen had received the Mandate to "possess *Sifang*,"[129] and that he had such great power that "throughout the Four Quarters none dared to insult him"; "throughout the Four Quarters none dared to oppose him";[130] "in the Four Quarters men are influenced by his power"; and "all in the Four Quarters of the state render obedient homage."[131] This total subjugation of the formerly foreign and unpredictable *Sifang* reinforced the absolute dominance and superiority of the center.

Such political superiority of the ruling clan, in both Shang and Zhou ideology, was legitimized by the king's monopoly of ritual communication with the divine world, through an ancestral line that formed a vertical axis of the layered cosmos. Constance Cook has recently argued

128 *Sifang* remained an important concept in Zhou ideology – there are twenty-four usages of *Sifang* in bronze inscriptions found in Shirakawa's concordance, eighteen usages of *Sifang* in the *Book of Documents* according to Gu Jiegang's *Shangshu tongjian*, and twenty-one usages in the *Book of Poetry* according to the *Harvard Yenjing Concordance of Shijing*.

129 "Da ya" and "Huang yi" in the *Book of Poetry*; see Legge, 1963, vol. 4, p. 451.

130 Ibid., p. 455.

131 "Da ya" and "Yi" in the *Book of Poetry*. See Legge, 1963, vol. 4, p. 511.

that the Zhou inherited from the Shang this vertical dimension of the cosmology, in which the ancestral spirits traveled between "up and down" – Heaven, ritual ground, and tombs.[132] Even though no royal tomb of the Zhou has been excavated, and we do not know if the Zhou continued the *ya*-shaped tomb structure, the bronze inscriptions and transmitted texts quoted earlier provide ample evidence of this continued vision of a monopolized pivot of the cosmos. The Zhou conceived their ancestors, like the Shang ancestors, as living in Heaven on the left and right of the high god, ascending and descending among the layers of the cosmos. Both Shang and Zhou kings monopolized this pivot as their access to the divine world, with the king as the only one who could communicate with his own ancestors and who controlled the bronze sacrificial vessels.

This continuing cosmology located the center of the cosmos in the king's body, which became the pivot between the vertical axis of his ancestral line and the horizontal exercise of power in conquering the Four Quarters of the earthly realm. Both Shang and Zhou kings were preoccupied with occupying a specific region on earth that was seen as the center – the Shang calling itself the "Central Shang," for example, and Zhou's construction of Chengzhou on the central plain. But neither of them was fixed to a single capital as the geographic center. Keightley proposes that the Shang king displayed his power by frequent travel, hunting, and inspection of his domains, and Cook points out that the Zhou king too was constantly on the move among many ritual sites in different cities.[133] It was the king – his body, his monopolized access to the divine world, his ability to communicate between the upper and lower levels (*shang xia* 上下) – that constituted the centrality. This centrality in a layered cosmos legitimized the king's possession of the Four Quarters. It was manifested in the tales of the Zhou's founding kings and their power to extend both to high and low (*shang xia*) and to the *Sifang*. The best example of such a formulated tale is the inscription on the Shi Qiang *pan*:[134]

Accordant with antiquity was King Wen!
He first brought harmony to government.
The high god sent down fine virtue and great security.
Extending to the high and low, he joined the ten thousand states.
Capturing and controlling was King Wu!
[He] proceeded and campaigned throughout *Sifang*,
Piercing Yin (Shang) and governing its people.

132 Cook, 1997. 133 Keightley, 1983; and Cook, 1997.
134 "Shi Qiang *pan*" references are to Shirakawa, vol. 50, entry no. 15, p. 335. Translation is adapted from Shaughnessy, 1992, p. 185.

In this statement, King Wen's receiving the Mandate from the high god and his communication with the spiritual world – "high and low" – was the precondition for Zhou's possessing *Sifang*. Zhou's replacement of Shang was achieved in every cosmological dimension – becoming the center of the layered universe connecting the high and low, receiving the blessing from the high god, and joining the ten thousand states of *Sifang*.

Conclusion

The search for the origin of *Wuxing* cosmology in the Shang and Western Zhou periods has been attempted by many scholars. However, the issue remains ill-defined for our understanding of either ancient Chinese history or its cosmology. This is because the definition of *Wuxing* cosmology, either consciously made or taken for granted, has often been based on textual representations two thousand years after the Shang period. Because such a search looks for origins rather than transformations, it ignores the fact that similar symbolic forms, such as the classification of time and space, could have different meanings and functions in different contexts.

This chapter does not answer the question of the origin of *Wuxing* cosmology. Rather, it demonstrates its heritage from the Bronze Age – the Shang and Western Zhou. This heritage is far more essential to an understanding of correlative cosmology than is the origin of terms associated with *Wuxing*, that is, the function and meaning of the Shang cosmology in its own context. From this contextual study we can conclude that the *Sifang*-center cosmology of Shang China was a three-dimensional spatial and temporal structure. It functioned as a cultural totality in which both political interactions among human beings and ritual communication with the divine world were carried out. With this single structure of time and space, the ancient Chinese scheduled their political, economic, and ritual actions, built their cities, temples, and tombs, constructed calendars and geography, conceptualized space and time, and classified all the events, forces, and beings in the universe. The center of this structure was crucial both cosmologically and politically, and occupation of this central location itself constituted divine authority and political power. When the Western Zhou replaced Shang as the center, they inherited this cosmology and reinforced the political dominance of the center through a more efficient system of monarchy. As a cultural totality, *Sifang*-center cosmology encompassed many dimensions of life, and thus had multiple layers of meaning.

It is these multiple dimensions and layers of meaning of the *Sifang*-

center cosmology that became the core of correlative cosmologies – including that of *Wuxing* – during the imperial era. These dimensions include the political classification of the Four Quarters as the "others or subordinate" and the ruling clan as "us"; the geographical division between the four lands and the "central Shang" or the "central region" of the Western Zhou; the spatial conception of the four directions and the center; the division of time into the seasons, months, years, and time periods of the day; the architectural structure of four sides with a center; and the classification of forces and beings in the universe in terms of four *fang* and the center. It is the layering of these dimensions in the single *Sifang*-center structure that became the foundation for the correlation-making on a grand scale of later times.

The fundamental transformation of *Sifang* cosmology into the *Wuxing* system occurred during the transition between two historical ages, the Bronze Age and imperial China, that happened from the fifth to the second centuries B.C., the period to which Chapter 3 is devoted.

3

Wuxing: Cosmology in Historical Transition

Introduction

The Warring States period, from the fifth century B.C. to 221 B.C., was a time of disarray in the long historical transition between two political ages, from the Bronze Age, dominated by the ruling clans of Shang and then Zhou, to the creation of the first unified empire by the Qin. In this transition, changes in cosmology and in sociopolitical relations constituted a single historical process, which reconstructed social relations as well as knowledge of the universe.

During this period, social and political relations went through unprecedented comprehensive changes. The Zhou court was first driven from its capital by invaders from the West in 771 B.C. Eventually its lingering symbolic existence was ended at the hands of rising local rulers who were previous Zhou lords and nobility. The wars for new hegemony (*ba* 霸) were constantly being waged between states, while within each state, civil wars frequently broke out when the rising ministers or military leaders overturned the hereditary lords of the former Zhou ruling clan. During such constant internal and external competition, new social ties created through blood covenants (*meng* 盟) formed a realm of non-kin political relations that broke the hierarchical and political order based on the Zhou lineage system.[1] To increase their military force, the states replaced the chariots of aristocracy with infantry armies composed of the common peasants. This extension of military service to the common peasantry, according to Mark Lewis, created new mechanisms of mass mobilization and control that soon were used to control

1 The archaeological excavations at Houma provided the most reliable and comprehensive evidence about the forms and functions of blood covenants during the late Spring and Autumn period. For publication of the texts of oaths that resulted from numerous blood covenants at Houma, see Shaxisheng wenwu gongzuodui weiyuanhui, 1976. For scholarship on Houma covenant texts, see Susan Weld, 1990. For a general discussion of blood covenants, see Mark E. Lewis, 1990, pp. 43–52.

the civil population as a whole, and thereby changed the basic units and bounds of society. These control mechanisms included taxation on land, registration of the population, re-ranking the population according to military merit, growing bureaucratic administration, and the absorbtion of the defeated states into "districts 縣" of the central government of the territorial state.[2] With the growth of local administration, a class of civil servants emerged, seeking employment from the new autocratic rulers of the territorial states. In the end, a social hierarchy based on the lineage system of the royal clan was replaced by territorial states administered by a body of salaried civil servants. The hereditary political power of the king, which was sanctioned by ancestor worship, gave way to rulership sanctioned by a new cosmology that directly related human sovereignty to Heaven and its cosmic pattern.

There is an intrinsic and constructive component of the institutional changes outlined here; that is, the creation of a cosmological discourse – a symbolic system that depicts the order of the cosmos. Historians have long focused on the institutional changes alone, overlooking this symbolic system. Yet some scholars have started to incorporate cosmology into their analyses. Robin Yates has recently argued that cosmology as a symbolic system formed a powerful discourse which appeared to be "self-evident and true," and through which both intellectuals and ordinary people spoke and thought.[3] Mark Lewis also emphasizes that there was a "reciprocal dialectic" between cosmology and the new political order during the Warring States period.[4] Both Yates and Lewis confirm that cosmology was a constructive part of institutional change – it was invented to sanction the new order and institutions,[5] and it created boundaries of time, space, and body that provided a structure for the daily practice of the bureaucracy of the Qin state.[6]

It is generally accepted that correlative cosmology as a political discourse was a Warring States creation, a creation that has been conventionally studied by investigating the philosophical texts of the time. Little attention has been paid, however, to the larger cultural and historical foundations of this creation. This chapter demonstrates how this cosmological discourse developed from, and in turn transformed, the previous cosmology of *Sifang*. It also shows that this transformation was intrinsic to the construction of new power relations and the destruction of political relations that had persisted through the two millennia of the

2 Lewis, 1990. For examination of military organization, the development of bureaucracy, and social and political change during the Warring States period, see also Robin Yates, 1987; 1988; 1994a; 1995.

3 Yates, 1994a, p. 56. 4 Lewis, 1990, pp. 213–41. 5 Ibid., p. 12.

6 Yates, 1994a, p. 57.

Bronze Age. To investigate Warring States cosmology as such a transformation, this chapter traces the process of remaking cosmology from the *Sifang* structure to the correlative system of *Wuxing*. It outlines three major ways in which the transformation of cosmology implemented the political change.

First, the flourishing of multiple correlative systems, including *Wuxing*, implemented the separation of divine knowledge from kingship – the king's body and his ancestral line. These correlative systems formed a body of knowledge that was no longer monopolized by the king, but instead was possessed and reproduced by the rising political groups – including religious and natural experts, the rising ministers and bureaucratic officials, military professionals, and the emerging cultural elite or scholars. By using correlative cosmology as a discourse for political argument, as a source of ultimate authority for their professions, and as guidance for their daily practices, these groups usurped the hereditary king's monopolized divine authority and distributed it among themselves, eventually changing the nature of rulership altogether.

Second, the "center" of the cosmos in *Sifang* cosmology had been the king and his clan, who had supreme domination over the four *fang*, and this superiority of the center had reached a climax with the Western Zhou monarchy. The reduction of this center to the fifth *fang*, due to the collapse of the Zhou central power and the rise of local powers, marks the fundamental transformation of the cosmology. By transforming the *Sifang* structure depicting a sacred center into a dynamic system of five interactive phases (*Wuxing*), the rising political groups undermined the absolute and sacred centrality of the former ruling clan, replacing it with the dynamic conquest and generation characteristic of the *Wuxing* system. Such a dynamic cosmology provided a logical pattern and rationale for social change, forming a symbolic structure for power competition and ratifying the drastic change taking place in power relations.

Finally, the new correlative cosmology redefined the concept of rulership by creating a direct connection between Heaven and Man that denied the king's monopolized access to the divine world through his ancestral line. Originating from multiple kinds of correlative categories – those based on binary pairs, triads, quadruplets, or quintuplets – the correlative cosmologies were gradually developed into a single coherent correlative cosmology. This synthesized cosmology formed a common discourse for debate among various political forces, for sanctioning as well as regulating the new political rulers, and for ordering the daily practices of the populace.

It is my central argument, therefore, that the emerging cosmological discourse, with the *Wuxing* system at its core, was fashioned by the political actors of the time in order to dismantle the previous power relations, to reshape the connection between political power and divine knowledge, to implement changes in social and political institutions, and to dispute the patterns of history leading to the new social order.

Knowledge and Authority beyond the Kingship

The Usurpers of the King's Divine Authority

During this period of drastic social change, political disintegration transferred the business of connecting Heaven and Earth from the king to various layers of society – to the courts of warlords, to ritual specialists and ministers, and even to bureaucratic officials and diviners in local communities. This separation of the political ruler from divine knowledge, which was accompanied by breaking of the hereditary rulers' previous monopoly on access to Heaven and the spirits, began as early as the end of the Western Zhou. Constance Cook describes how, after the fall of the Western Zhou hegemony, the ritualists as a rising class robbed the king of his position as the central pillar of communication with Heaven and spirits, forcing him to depend on the bureaucracy of the ritualists for his connection to Heaven, and eventually breaking that connection.[7]

By the third century B.C., four major social groups had ascended to grasp access to divine knowledge and authority – that is, communication between Heaven and Man. First, there were religious and natural experts.[8] Long ignored by historians, who have emphasized instead predominance of the philosophical traditions of this period, this group has drawn scholarly attention only in the past decade because of new archaeological discoveries that have demonstrated their overwhelming influence in the culture of the time. Religious and natural experts included a great variety of specialists – ritualists, diviners, astrologers, musicians, physicians, calendar makers, and others who served the new patrons of local power with their expertise. A. C. Graham has attributed the development of correlative cosmology – that of *Yin* and *Yang* 陰陽, *Wuxing*, and *qi* 氣 – precisely to this group of religious and

7 Cook, 1995.

8 I adopt the term "religious and natural experts" as an analytical category, rather than as a native category of the Warring States period, from Donald Harper, 1999. The retrospective labels for this group used by the Han Chinese include *Yin-Yang jia* 陰陽家 (*Yin-Yang* specialists), and *fangshi* 方士 (recipe gentlemen).

natural experts who were outside the philosophical tradition of the time.[9] Donald Harper sees this group as "vital participants" in the intellectual and spiritual ferment of the period.[10] Li Ling further claims that the knowledge developed by these experts – "*shushu fangji* 數術方技," in Chinese terminology – preceded and then paralleled the philosophical tradition, creating dual trends in Chinese culture.[11] In their profession, these experts started to build various correlative classifications between the natural order and the human world based on numerological sets of two, three, four, five, and so on. By applying these correlative categories to their own specialties, while serving their patrons, they replaced the king's monopoly of communication with the divine with their own innovation – direct correlation of Heaven, Earth, and Man.

The second group who usurped the power of communication with the divine were the rising bureaucratic officials and local administrators. Though not credited as the innovators of correlative cosmology, the bureaucratic class was the most powerful in applying correlative cosmology to implement and control the new political order. According to Cook's study, by the third century B.C. the Chu bureaucrats not only had become the clients of diviners and ritualists, they themselves were directly engaged in Heaven and Man communication and thus had "completely usurped" religious functions originally associated with the tradition of ritualists in the Zhou court.[12] With the extension of bureaucratic institutions into the local society, local administrators also used cosmological texts, such as almanacs and various diagrams and charts for mantic practices, as guides to the practice of local administration. The archaeological discovery of the tomb of a Qin local official named Xi 喜, who died in 217 B.C., reveals such mantic practices in local administration. Robin Yates has demonstrated that, since the bureaucratic procedures of the Qin state were closely linked to the cosmos or divine, the use of such cosmological guidance by local officials became an integral part of administration. Thus in the daily function of the bureaucracy, the bureaucrats, from ministers to local officials, occupied a "central and essential position in the regulation and control of the entire cosmic system."[13]

The third group, the emerging military specialists, attacked the hereditary nobility by transforming warfare from a form of service to the ancestral cults to an expression of cosmological patterns. Mark Lewis describes

9 Graham, 1986, pp. 91–2.

10 Harper defines this group as "natural experts," who were responsible for the creation of correlative cosmology and at the same time embraced the magical religious elements of the culture of the time. See Harper, 1999, p. 817.

11 Li Ling, 1993, pp. 2–18. 12 Cook, 1995. 13 Yates, 1995, p. 333; 1994, p. 79.

how such military specialists invented military treatises as expressions of "divine patterns inherent in the cosmos" and used such a "textual vision of warfare" to deny the chivalric warfare that had served the ancestral cults – the very basis of the hereditary power of the nobility.[14]

The last group who usurped the hereditary ruler's divine authority were the scholars – masters and students engaged in private learning and seeking employment at various level as advisors, educators, and cival servants. They elevated the assault on monopolized divine access by the hereditary ruling clan to a theoretical level, advocating the superiority of achieving connection with the divine through education. William E. Savage states that such diffusion of Bronze Age theology was expressed in the Confucian *Analects*, which advocated the superiority of achieved excellence by the Confucian "gentleman (*junzi* 君子)," whose connection to the divine and right to rule were the result of education rather than of inheritance.[15] A. C. Graham points out that, although not the innovators of the correlative cosmology, the philosophic schools in the third century B.C. adopted the correlative cosmology in their literature.[16] It was the scholars, moreover, who eventually accomplished its synthesis into a coherent system.

These four political groups are analytical categories, retrospectively imposed to assist discussion. They are not meant to imply that the Warring States political actors had such group identities, nor that their social identities coincided with these four analytical categories. Religious and natural experts, for example, were identified at the time by their specific professions, such as scribe (*shi* 史) or physician (*yi* 醫); similarly, bureaucratic officials were identified by their ranks. A Warring States category like *shi* 士, by contrast, could be applied to various people across all four groups.[17]

These rising political groups used two major methods to deprive the hereditary ruler of monopolized divine authority and to disperse such authority among themselves in order to increase their own power. One method was converting the cosmology and ritual practices of the Bronze Age into something new in order to serve their own ends. A good example of such a conversion is the covenant feast. The ritualists converted the Zhou gift-giving feast, which originally functioned to link the living king with his ancestors in Heaven, to a covenant feast that actu-

14 Lewis, 1990, pp. 97–114. 15 Savage, 1992.
16 Graham, 1986, pp. 91–2; 1989, pp. 313–82.
17 The nature of the *shi* class in Chinese history is best analyzed by Yu Yingshi. Yu points out that during the Spring and Autumn and Warring States periods, *shi* became a melting pot where the declining aristocrats and rising commoners merged. He defines *shi* in these periods as intellectuals – people who possessed knowledge and aimed at realizing the human order through government. Yu, 1987, pp. 12, 26–51.

ally broke the political order based on ancestor worship and created a realm of non-kin political relations.[18] The other method was the invention of numerous mantic techniques of communicating with the divine world.[19] The two methods were often inseparable in practice, since the new systems of knowledge often transformed and integrated traditional concepts and practices. In this process of transformation and innovation, the traditional mantic techniques used by the court, such as divination, oracles, and astronomy, acquired new patrons and performed different functions. At the same time, new techniques flourished – the reading of omens, the consulting of almanacs, and military strategy. While diverse, most of these mantic practices gradually incorporated the correlative categories that build direct connections between all things in the universe. Therefore, these transformed or invented means of access to the divine world together came to form a highly complex, multifaceted, and fluid cosmological discourse, which was indispensable for political action and social practice.

Wuxing, one such emerging cosmology, circulated in most of the mantic practices. It best exemplifies how such a correlative scheme portrayed the cosmos as an ordered system that lay behind historical events and beyond the ruler's manipulation. *Wuxing* was used in a variety of mantic or professional practices, providing a discourse for political discussion and competition, for maximizing personal success by following the natural rhythms of the cosmos, and for ordering the daily existence of the populace at various levels of society.

In this discourse the voices of various political actors, no longer muffled by that of the king, were heard; and an authority of cosmic order, no longer monopolized by kingship, emerged. The following section examines three levels of social practice at which a *Wuxing* discourse was employed: those of state affairs, military command, and the daily functions of local administrations and communities.

Wuxing in State Affairs

First, *Wuxing* was used at the state level in political discussions. *Zuozhuan* 左傳, a historical chronicle dating from the fifth to fourth century B.C.

18 For the rise of the covenant feast (or blood covenant), see Lewis, 1990, pp. 43–52; and Weld, 1990. For a detailed study of how the covenant feast actually evolved out of early Zhou gift-giving and mortuary ceremonies, both crucial to the ancestral cult, see Cook, 1995.

19 Michael Loewe distinguishes three categories within mantic practices. Divination results from deliberate steps taken by man to induce signs. Oracles and omens are signs inherent in nature, with the former reflecting the regular order of nature and the latter conflicting with it. See Loewe, 1982, p. 91; and 1986c, pp. 673, 679.

and later attached to *Chunqiu* 春秋 or *Spring and Autumn* as a commentary,[20] demonstrates such uses of *Wuxing* in the last quarter of the text, which comments on the years of Dukes Zhao, Ding, and Ai. Although not a precise history of the *Chunqiu* period, *Zuozhuan* reveals its practices up to the time of the fourth century B.C. It describes, in fragments, the use of an astrological system in political discussions – a field allocation (*fenye* 分野) system used to predict the course of events based on celestial and terrestrial correspondences. This astrological system used Fire, Water, and Metal to correlate polities on earth with heavenly bodies. This rudimentary form of *Wuxing* was used in discussions of the pressing political issues of the time.

Most commonly, court diviners or astrologers used the field allocation system to explain changing power relations among states. In the ninth year of Duke Zhao (533 B.C.), for example, there was a fire in the state of Chen. Based on this fire, a diviner of Zheng, named Pi Zao, predicted the reestablishment of Chen and repulsion of Chu five years later, and the final fall of Chen fifty-two years after the reestablishment. He made this prediction according to the correlation between the states and celestial bodies categorized by Water and Fire in the field allocation system:[21]

> Chen belongs to Water. Fire is antagonistic (*fei* 妃) to Water and is under the regulation of Chu. Now the Fire [the star named *Huo* 火 or Great Fire 大火, Sco α] has appeared and kindled this fire in Chen, [indicating] the expulsion of Chu and the establishment of Chen. Antagonists are ruled by the number five [in their conjunctions], and therefore I say in five years. The Year star must five times come to *Chunhuo* 鶉火 [constellation *Liu* 柳, Hya ∂], and then Chen will finally perish . . . This is the way of Heaven, and therefore I said fifty-two years.

Whether this prediction had any real influence on policy making is unknown. But this case reveals a new relation between rulership and knowledge. By attributing a shift in state power to a cosmic order – that is, Water and Fire ruled by the number five – this diviner claimed that his special knowledge gave access to the hidden pattern behind political events, a source of divine support outside the ruler's grasp. The ruler was transformed from the monopolizer of divine wisdom to the patron of specialists and knowledgeable men. A ruler's power partially depended upon his having access to resources of such knowledge.

The cosmological discourse was not limited to professional diviners, however. Statesmen also used this discourse to discuss state affairs. In

20 For discussion of *Zuozhuan*'s authorship and date, see Ann Cheng, 1993.
21 *Zuozhuan*, Zhao 9.

the seventeenth year of Duke Zhao (523 B.C.), for example, a comet appeared to the west of *Da Chen* 大辰 [another name for the star *Huo* 火 or Great Fire 大火, Sco α] and traveled to the Milky Way (which belonged to Water). Accordingly, two ministers of Lu, Shen Xu and Zi Shen, forecast calamities of fire. Zi Shen further predicted that four specific states would have fires on certain days:[22]

> The four states to which this comet has reference will be, I think, Song, Wei, Chen, and Zheng. Song is the region corresponding to *Da Chen* 大辰 (Sco α); Chen was the old abode of Da Hao; Zheng, that of Zhu Rong – all of them abodes of Fire. The comet is traveling to the Han [the Milky Way] of the sky, and the Han is ominous of Water [to which Zheng corresponds]. Water is the husband [*mu* 牡] of Fire. The calamity will arise on *bingzi* day or *yinwu* day, when there is a meeting of Water and Fire.

According to *Zuozhuan*, the four states did have fires on the predicted days.

Yet the ruler was by no means always manipulated using knowledge of this kind. This prediction only served to open a political debate in the court of Zheng about how to run the state and who should run it.[23] Following the prediction of the catastrophe of fire, diviner Pi Zao, supported by the majority of the court, claimed that if he could use state treasure to perform rituals, Zheng could avoid the catastrophe of fire. Minister Zi Chan, competing with the religious expert for authority, rejected Pi Zao's demand by claiming that "the Way of Heaven is far away" and beyond the reach of man. Dismissing the diviner's knowledge as mere "talk," Zi Chan took total control of the situation. This debate became a real power contest, in which the cosmological discourse was used in the competition among different political factions – the religious experts versus the ministers. As seen in this case, not only were rulership and knowledge separated, but power itself was composed of diverse factions. A weak ruler could be manipulated by the strongest faction, while an active ruler had to maximize his power by forging consensus among different groups and inspiring a collective effort.[24]

One good example of cosmological discourse for political discussion is the case of the ruler of Jin, Zhao Yang. In the ninth year of Duke Ai (486 B.C.) the ruler of Jin consulted the tortoise shell about military assistance for Zheng against Song, and received the indication of Fire meeting Water. He asked three religious experts titled scribes (*shi* 史) –

22 *Zuozhuan*, Zhao 17.

23 The idea that divination and interpretation of omens were means of political discussion has been greatly inspired by correspondence with Professor Nathan Sivin.

24 For a discussion of the ideal ruler in the theories of the Warring States period, see Michael Nylan, 1992, pp. 14–44.

Zhao, Mo, and Gui – for their interpretations of the divination. Zhao, Mo, and Gui each interpreted the sign according to his own special knowledge, but all agreed that Jin should not attack Song to aid Zheng. According to the field allocation system, Mo specifically explained why Jin should not attack Song but should attack Qi:[25]

> Ying [the surname of the Jin ruler] is a name of Water, Zi [the surname of the ruler of the Song] is the position of Water. To put the name and the position in antagonism is not to be attempted. Yan Di had a Fire master, from whom the house of Jiang is descended. Water conquers Fire. According to this you may attack the states with the surname of Jiang.

In this case, the Jin ruler gave up the attempt to attack Song and a year later attacked Qi, whose surname corresponded to Fire. It is not clear if this decision was made because of the interpretation of the divination. But once he had decided to invade, the ruler of Jin used these interpretations to justify his decision when challenged by his ministers.[26]

In the astrological system as quoted here, three of five symbols of *Wuxing* are mentioned. Besides Water and Fire, scribe Mo used "Fire conquers Metal" in his interpretation of the solar eclipse that occurred in the thirty-first year of Duke Zhao (509 B.C.), predicting that six years later the state of Wu would invade the Chu capital Ying 郢 but would not succeed. Other uses of *Wuxing* include the names of the offices of *Wuxing* 五行之官 – the offices of Wood, Fire, Metal, Water, and Earth, which were said to be rooted in remote antiquity – as stated by a Jin minister, Cai Mo;[27] the grouping of Water, Fire, Metal, Wood, and Earth together with Grain to form "the Six Stores" (*liu fu* 六府) by a Jin official;[28] the attribution to Heaven of the birth of the Five Materials (*wu cai* 五材) by a minister of Song, Zi Han;[29] and the attribution of *Wuxing* as belonging to Earth – "地有五行" – by scribe Mo of Jin.[30]

Although represented in such a fragmented and incoherent fashion, the use of five *xing* in *Zuozhuan* had elementary characteristics of the later, systematized version of *Wuxing*: Water, Fire, and Metal in the field allocation system were related in cycles of antagonism, meeting, or conquest; and such cycles combined spatial correlation between states and constellations with a temporal pattern of cosmic movement into a cosmology, explaining unpredictable changes in the human world according to a predictable cosmic order. For the political actors, to move with the fluid cosmic movement rather than against it was essential to their survival and success.

25 *Zuozhuan*, Ai 9. 26 *Zuozhuan*, Ai 10. 27 *Zuozhuan*, Zhao 29.
28 *Zuozhuan*, Wen 7. 29 *Zuozhuan*, Xiang 27. 30 *Zuozhuan*, Zhao 32.

The authority of the cosmological discourse thus lay outside of kingship, in a permanent cosmic pattern accessible to various political groups, and no longer accessible exclusively through the king's connection to his ancestors. Major state affairs – military alliances with certain states, assaults upon other states, prediction and prevention of catastrophes – were no longer determined by the ruler through the ancestral wisdom he received from his monopolized ritual communication, but had to be debated and discussed by various political groups – such as religious and natural experts and ministers – based on the cosmological systems accessible to their respective professions.

Wuxing *in Military Treatises*

The cosmological discourse provided a new source of authority that lay above and beyond the direct control of a political ruler. This authority was appealed to not only by religious experts and ministers at the court in political discussion, but also by the emerging military professionals in their formulation of military treatises – the text-based authority of their profession. Lewis has demonstrated that such theoretical formulations of warfare were major elements in redefining the nature of authority and the state. The military treatises rejected the old Zhou aristocracy by divorcing warfare from ancestral cults, and they helped shape the new modes of political authority by applying new models and techniques of military command to the creation of a new bureaucratic state apparatus.[31] The authority of such military professionals and their texts, after all, lay in discerning hidden patterns of the cosmos and in using them to create order, in warfare as well as in the new society.

The cosmological patterns, the correlation between seemingly chaotic human affairs and the hidden cosmic order, constituted the major source of authority in military treatises, and *Wuxing* was one of the many systems used in depicting such patterns. A long-lost military treatise, *Sun Bin bingfa* 孫臏兵法, dated to the middle of the Warring States period, was unearthed in 1972. It depicts a conquest cycle based on five soils of different colors: "Green conquers Yellow, Yellow Black, Black Red, Red White, White Green."[32] According to the correlation of these colors with *Wuxing* as represented in other sources, this cycle is identical to the conquest cycle of *Wuxing*. This conquest cycle of five soils thus depicted hidden cosmic patterns behind the seemingly random locations. A com-

31 Lewis, 1990, pp. 11, 97–135.
32 Yinqueshan Hanmu zhujian zhengli xiaozu, 1975, p. 61. For bibliographical information regarding military texts of the period, see Gawlikowski and Loewe, 1993, pp. 446–55.

mander's comprehension of such cosmic patterns ensured his victory in combat.

Besides structuring spatial order as a conquest cycle of five soils, *Wuxing* was also used to structure time as a recession and accretion of cosmic cycles. A grasp of this concept was essential for a commander if he was to gauge decisive timing for military success. A mantic practice called "*Xingde* 刑德" (recession and accretion) was specially dedicated to depicting such an order of timing in military action.[33] There are three versions of a "Xingde" text discovered in Mawangdui 馬王堆, which was buried by 168 B.C.[34] This text manifests the shifts of cosmic power among five "powers (*de* 德)" and dictates military actions accordingly. The phase of cosmic power in which a military action was taken determined its consequences:[35]

> The cosmic power (*de*) is at Wood . . . Taking a military action at this moment, the heart of the population will be exhausted, gentlemen will wear armor to the court and little men (*xiaoren* 小人) will flee the land with their wives . . . The party that initiated the warfare will lose territory and weaken its troops.

When the cosmic power is at Metal, which conquers Wood, the consequence of the action is the opposite: it is the proper time for military campaigns, and the ones who initiate the warfare will win, while the ones who join later will perish.

The *Wuxing* system, as seen in the "five colors of soil" and the "five powers," was adopted in military treatises as an art of command; it was used by army commanders in choosing the moment and position in battle to ensure victory. By grasping the cosmic pattern as such, the military professionals denied the value of the old aristocratic warfare that was part of ancestor worship, replacing the king's monopolized ancestral wisdom with a knowledge of cosmic patterns, now accessible through military treatises, as the authority for conducting combat and commanding men.

Wuxing *in Bureaucracy and the Daily Life of the Populace*

While the correlative cosmologies, as illustrated by the rudimentary *Wuxing* system, were used by various political groups at the court in

33 "*Xingde*" used to be translated as "punishment and reward." John Major has persuasively argued that "punishment and reward" is only suitable when referring to an overt action of human agency. When referring to a cosmological principle, as in military treatises, it should be translated as "recession and accretion" – an indication of the cosmic cycles such as the seasons, the Five Phases, or *Yin-Yang* forces. See Major, 1987.

34 Fu Juyou and Chen Songchang, 1992, pp. 132–43.

35 Ibid., pp. 136–7.

redefining the nature of political authority, they were also indispensable in formulating the bureaucratic system and administering the vastly diverse populace, preparing institutionally and culturally for unification under a single centralized empire.

Yates's recent studies on the Qin bureaucratic system reveal this long overlooked cosmological foundation of the birth of China's bureaucracy. Yates asserts that early Chinese bureaucracy "was conceived, first and foremost, as based on the principles underlying the functioning of the cosmos."[36] The primary evidence for such a cosmological foundation of early bureaucracy, particularly for the day-to-day practice of local administration, consists of two calendrical texts, called almanacs or *rishu* 日書, discovered in the tomb of a local official of the Qin named Xi 喜, who died in 217 B.C.[37] These two almanacs were accompanied by many legal documents buried in the same tomb. Like those legal documents, the almanacs were practical guides to be consulted in the daily functioning of local administration. The most direct guidance for local officials in these almanacs is found in two texts entitled "Officials" (*li* 吏).[38] One of the texts, written on the upper five registers of slips 886–895 of the almanacs, contains prognostications of the results of interviews between a lower official and his superior. They range from positive results such as receiving praise (*you mei yan* 有美言) or gaining an appointment (*qing ming xu* 請命許), to negative results such as receiving evil words (*you e yan* 有惡言), not being able to speak up or be listened to (*you gao bu ting* 有告不聽), or having nothing accomplished (*baishi bu cheng* 百事不成). The other text, written on the lowest register of the same slips, prescribes the auspicious days on which to take up office (入官良日). Both texts are based on the twelve earthly branches, and the prognostications on the upper registers further divide the day into five periods of time.

In addition to these direct guidelines for an official's activities, the almanacs equipped local administrators with the routines of communal activities – rituals, trade, farming, and so on – and responded to the social problems of slaves and farmers fleeing the land as a result of both changing class relations and newly enforced tax policies. Such guidance

36 Yates, 1995, p. 333. See also Yates, 1994a.

37 These two calendrical texts, called *Rishu*, were discovered at Shuihudi, Yunmeng, Hubei province, in 1975, and were published first in an excavation report in 1981, *Yunmeng Shuihudi Qinmu*, pp. 69–70. They were published with transcription and commentary in 1990 by Shuhudi Qimu zhujian zhengli xiaozu in *Shuihudi Qinmu zhujian.* For major studies of these texts, see Rao Zongyi and Zeng Xiantong, 1982; Kalinowski, 1986; Loewe, 1988; Poo Mu-chou, 1992; 1998.

38 For publication of these texts, see Shuihudi Qimu zhujian zhengli xiaozu, 1990. For the study of these two particular texts, see Lin Jianming, 1991; Yates, 1995, pp. 339–42.

to local administration can be illustrated by two specific texts that aided the newly established legal system in identifying and capturing thieves. One of the texts is titled "Thieves 盜者," written on strips 827b–814b of the almanacs. It gives predictions for the appearance of thieves, the places they hide, and the best times of day to capture them successfully. The predictions are based on the twelve animals that represent the twelve earthly branches. The other text, entitled "Thieves 盜," written on strips 1148–1154 and 1154b, makes predictions about their gender, physical marks, and the directions in which they will flee.

The cosmological construction of the early bureaucracy coincided with the spread of correlative cosmologies in the mantic practices of the vast populace during the late Warring States period. Besides providing a foundation for the daily functioning of the state bureaucracy, correlative cosmology, as illustrated by the almanacs, also spread through the middle and lower layers of the society, becoming a common cultural phenomenon that paved the way for the unification of a vastly diverse populace. Almanacs such as those found in the local official Xi's tomb were also found in geographically distant regions, and were used by local diviners whose patrons were from the middle and lower layers of society. Besides the two almanacs found in Xi's tomb, two other Warring States calendrical texts were found in the tomb of a person named Dan 丹. According to the burial, Dan was likely an important person with some magical power in the local community. In this burial, the almanacs were the only documents found, except for a short record of the life of Dan and his magical power.[39]

These four almanacs reflect calendrical practices and a worldview commonly encountered among the middle and lower levels of society in diverse localities in the third century B.C.[40] While the almanacs buried in Xi's tomb provided guidance for the practices of the local bureaucracy, the range of concerns in these almanacs goes far beyond the bureaucratic system, covering the daily activities of these middle and lower layers of the society.[41] The almanacs addressed not only local officials, but also landowners, farmers, craftsman, and traders, prescribing a wide range of their daily activities. It may be argued that it was precisely because of their deep roots among the populace that cosmology

39 These two almanacs were unearthed at Fangmatan, Tianshui, Gansu province. For an excavation report, see Gansusheng wenwu kaogu yanjiuso, 1989. For publication and studies, see Qin jian zhengli xiaozu, 1989, pp. 1–6; and He Shuangquan, 1989.

40 Poo Mu-chou argues that the almanacs were spread in the "subculture" of the time; their users were mainly among the middle and lower layers of the population. See Poo Mu-chou, 1992.

41 For a detailed analysis of life and beliefs among the middle and lower layers of the society reflected in the almanacs, see Poo Mu-chou, 1992; 1998.

and calendrical texts had to be taken seriously by the local governing bureaucrats.

The four discovered almanacs concern daily activities of the populace, including performing rituals, childbirth, marriage, digging wells, building houses, making food, wine, clothing, and vehicles, buying cattle, grain, and servants, trading at market, traveling, and hunting. In addition, specific activities and situations are discussed separately in detail, such as sickness, dreams, death, detecting thieves, planting certain kinds of crops, recognizing and dispelling demons, and even the dates and directions for slaves to flee in exile.

These almanacs used many correlative systems, including *Wuxing*, to classify the days of a year into auspicious, inauspicious, and prohibited days for these various activities. Accordingly, the success or failure of all daily activities and the destiny of each individual were determined by the temporal and spatial orders of cosmic movement. The format of an almanac was an assemblage of various charts, diagrams, and prescriptions for mantic practices, in which *Wuxing* was one of many systems used. In these almanacs, *Wuxing* takes a more mature form. A complete set of five *xing* is arranged in both conquest and generation cycles and has many correlations, including day-signs, the months, seasons, directions, colors, animals, spirits or spells, sickness, and musical notes.[42] In short, almanacs, and *Wuxing* as used in them, addressed the conflicts and changes confronted by local communities during this period of drastic social change, and became a source of guidance for local communities.

Such calendrical practices show that as a new discourse, correlative cosmology, and *Wuxing* in particular, spread not only through these different layers of society, but also through different regions. Evidence of the use of *Wuxing* in calendrical practices has been found in the three regions of greatest political importance and cultural diversity at the time of unification – Qi, Chu, and Qin. Two almanacs found in Dan's tomb were buried by 239 B.C. in a northwest region of the Qin state before the unification of the Qin empire.[43] Two others found in the tomb of Xi were buried by 217 B.C. on Chu territory newly unified by Qin.[44] And Qi not only is famous for being the home of the believed innovator of *Wuxing*, Zou Yan, but also is known to have employed calendrical practices, as shown in the text of *Mozi* 墨子: "Mozi was traveling north toward Qi and ran into a calendrical diviner. The diviner said: 'The god is killing

42 Shuihudi Qimu zhujian zhengli xiaozu, 1990: "Rishu A," strips 68a2–77a2, 95a3–98a3; "Rishu B," strips 79b2–87b2, 189a1–195a1; transcription and commentary on pp. 193, 195, 239, 247.

43 He Shuangquan, 1989.

44 Shuihudi Qimu zhujian zhengli xiaozu, 1990, p. 2.

the black dragon in the Northern *fang*, your master's color is black, so you must not travel northward.' "[45]

The use of *Wuxing* in mantic practices continued during imperial times. In the Han Dynasty, *Wuxing* was elevated to become one of the most influential systems of mantic practice. Many mantic texts of the Han Dynasty reflect the spread of *Wuxing* cosmology. The popularity of *Wuxing* in "popular culture" did not, however, diminish its success in the imperial court. In order to decide on a day for taking women, Emperor Wu consulted diviners of varying expertise and got seven different answers. He decided to follow the prediction according to the *Wuxing* system since it was the predominant system.[46]

All the evidence discussed here shows that *Wuxing* had become a common discourse across class and regional boundaries by the third century B.C. The use of *Wuxing* at multiple levels of society, however, does not suggest a coherent system accepted by the entire population. Nor does it confirm a linear evolution of the idea. On the contrary, the development of *Wuxing* constantly manifested flexibility in structure and multivalence in meaning. First of all, *Wuxing* in the Warring States period was by no means the predominant cosmological system. It was often used in conjunction with other classification or number systems, complementing, integrating, or competing with them. These systems could be based on four, six, eight, or twelve. For example, the almanacs from Shuihudi 睡虎地 have a complete set of *Wuxing* correlated to the five colors, the five directions (*Sifang* plus the center), and the ten heavenly branches of day-signs. But when combined with *Sifang* or four directions, four seasons, twelve months, and twelve-branch day-signs, the correlative system often shifts to the base of four, with correlations identical to those based on five but omitting Earth.[47]

Second, the name *Wuxing* referred to different systems; and the same system of Water, Fire, Wood, Metal, and Earth was known by different names. In *Zuozhuan* and "Hongfan 洪範," *Wuxing* is the name for Water, Fire, Wood, Metal, and Earth; but in other texts, it refers to five kinds of moral conduct – benevolence, wisdom, rightness, ritual, and sagehood – as the teaching of Zi Si 子思 and Mencius 孟子 found in *Mengzi* 7, in a Mawangdui silk manuscript, and in a Guodian 郭店 bamboo document.[48] And the set of Water, Fire, Wood, Metal, and Earth, besides

45 *Mozi*, vol. 12. 46 *SJ* 127, p. 3222; *SJHK* 127, p. 14.

47 Shuhudi Qimu zhujian zhengli xiaozu, 1990: strips 68a2–77a2, transcription and commentary on p. 193; and strips 95a3–98a3, transcription and commentary on p. 195.

48 The use of *Wuxing* to refer to five kinds of moral conduct was first evidenced in a silk manuscript discovered in Mawangdui tomb no. 3, dated 206–195 B.C. (see "Laozi jiapian juanhou guyishu: *Wuxing*," in Guojia wenwuju guwenxian yanjiushi, 1980, vol. 1, pp. 17–28). A recent document discovered in tomb no. 1 at Guodian, also con-

being called *Wuxing*, was also called *wu cai* 五材 (Five Materials, as in *Zuozhuan*), *wu de* 五德 (Five Powers, as in *Lüshi chunqiu* 吕氏春秋), and *wu wei* 五位 (Five Positions, as in *Huainanzi* 淮南子). The incoherence in naming of the *Wuxing* system reveals that it remained very fluid while being formulated by multiple sources.

And finally, the function of *Wuxing* in building correlations also shows its multivalent nature. Diverse domains of the human world were incorporated into *Wuxing* separately to serve different interests. It was only through deliberate standardization by Qin and Han theorists that these correlations were integrated into a coherent total classification.

Wuxing, accompanied by other correlative cosmologies, emerged during the Warring States period. This was a significant step in the process of separating divine knowledge and authority from its old embodiment – kingship. During the Warring States period, the political ruler was no longer the center of the world, nor the monopolizer of access to divine knowledge. Correlative cosmologies, as exemplified by *Wuxing* – a new form of knowledge and a new source of authority – became a discourse shared by a diverse population. At court, different political factions, including the ruler himself, appealed to cosmology in arguing their cases. In warfare, military commanders used cosmology as the basis for the authority of their profession, claiming great autonomy from the ruler. In bureaucratic systems, officials consulted the cosmological patterns as guides to their daily functions in office. In local communities, diviners used them to advise patrons about their everyday lives. The old embodiment of power and cosmology in the king's body was separated into patrons and clients, into the territorial rulers and their hired servants. Knowledge of the divine powers and the cosmic order was no longer confined to the king's body and his ancestral line, but dispersed among divergent groups. These groups collectively usurped the hereditary king's authority, his connection to the divine. This process, as I shall demonstrate in the rest of the chapter, eventually transformed the nature of rulership from a hereditary king, legitimized by an ancestral cult, to a territorial ruler of a bureaucratic state, legitimized by his imitation of and conformity to the cosmos.

cerning the five kinds of moral conduct, is clearly entitled "*Wuxing*." The Guodian "Wuxing" document dates from the late middle Warring States period, that is, between 350–300 B.C. (see Jingmenshi bowuguan, 1998, pp. 29–36, 155–160). With these two pieces of evidence, scholars now commonly accept that this usage of *Wuxing* came from the teachings of Zi Si and Mencius, as suggested in *Xunzi*, ch. 6. A. C. Graham points out other examples of *Wuxing* as referring to five kinds of moral conduct (see Graham, 1986, p. 76). For discussion of *Wuxing* in the teachings of Zi Si and Mencius, see Pang Pu, 1981; Li Xueqin, 1998.

In this new relationship of knowledge and power, *Wuxing*, as a cosmic pattern behind all events and an authority independent of a specific ruler, became a resource for reorganizing social relations. The social and political changes in turn propelled the development of *Wuxing* and its transformation of the ancient *Sifang* cosmology.

From Centrality to Cycles of Change

While cosmological discourse was composed of multiple systems of correlation, the *Wuxing* system, later combined with *Yin-Yang*, eventually became the predominant system and the core of a synthesized correlative cosmology during the Qin and Han imperial era. One reason for such predominance of the *Wuxing* system, in my view, is that the political actors of the time found it most efficacious in integrating, challenging, and transforming the *Sifang* cosmology that was deeply rooted in the history of the civilization.

Being an intrinsic force of social change and political process, correlative cosmologies, *Wuxing* in particular, transformed the ancient *Sifang*-center cosmology and dismantled the previous mode of cosmology integrated with political power. First of all, *Wuxing* denied the concept of centrality in the *Sifang* cosmology, replacing the sacred and eternal center and its four subordinate *fang* with cycles of the Five Phases in which cosmic energies constantly interact and transform. Second, *Wuxing* changed the pattern according to which divine knowledge and political power were combined, substituting an all-encompassing system of direct correlation between Heaven and Man for the monopolized channel of communication with the divine – the king's body and his ancestral line.

It was in this process of transformation that *Wuxing* developed its two basic properties – cycles of dynamic interaction among cosmic energies and a scheme of correlation between the human world and the cosmos. Previous scholarship has traced the origins of both these properties of *Wuxing* either to theoretical innovation by individuals or to archaic symbolic sources. Graham concludes that the invention of generation cycles was only possible after Zou Yan had correlated the conquest cycles of the Five Powers with the colors and four seasons; while John Major traces the origin of conquest cycles to the more antique magic square.[49] While not denying the possibility of such individual or symbolic contributions to the origin of *Wuxing*, I believe that the basic properties of the *Wuxing* system emerged during a process of social practice and political change

49 See Graham 1986, pp. 81, 92; and Major, 1984, n. 17.

in the late Warring States period, and that it was during this process that new theories were created and cultural resources transformed.

Cycles of Cosmic Movement

In replacing the hierarchical structure of the *Sifang*-center system with cycles of interacting cosmic energies, *Wuxing* directly challenged the pre-existing political order. The astrological system of field allocation, as quoted earlier from *Zuozhuan*, shows that the direct correlation of each territorial state with cosmic forces denied the Shang and Western Zhou political landscape structured on four *fang* surrounding the center. Such celestial-territorial correlation directly defied the concept of an eternal and sacred center of both political and religious superiority. In contrast to the *Sifang* cosmology, which portrayed the hierarchical relationship between the four *fang* and the center as static and indisputable, *Wuxing* replaced the notion of an eternal centrality and hierarchy with constant interactions of cosmic energies. Water, Fire, and so on, together with their correlated states, interacted with one another in cycles of conquest and of meeting.[50]

While *Wuxing*, as used in the field allocation system, denied the concept of centrality and hierarchy in the *Sifang* cosmology, it did not discard the *Sifang* cosmology itself, but rather incorporated and totally transformed it by introducing cycles of cosmic movement into the previous spatial and temporal structures. In this recreated cosmology, *Sifang* as the four cardinal directions was grouped with the center, becoming five spatial units correlated to the five *xing*. These five spatial units existed in constant interaction in two major cycles of *Wuxing*, those of conquest (mutual overcoming) and generation (mutual production). Different versions of these cycles that developed during this period are listed in Table 3.1.

Among these cycles, the conquest cycle and the first generation cycle were the most popular. The conquest cycle is composed of: Wood conquers Earth (by tilling the earth with wooden plows), Earth conquers Water (by damming), Water conquers Fire (by extinguishing it), Fire conquers Metal (by melting it), and Metal conquers Wood (by cutting it). The first generation cycle is composed of: Wood generates Fire (by burning), Fire generates Earth (with ash), Earth generates Metal (from ores), Metal generates Water (by melting), and Water generates Wood (by irrigation).

50 Some scholars suggest that the generation cycles are indicated in *Zuozhuan*, especially in this case referring to "meeting cycles." See Li Hansan, 1981, pp. 31–2.

Table 3.1. *The cycles of* Wuxing *interactions*

Sequence of *Wuxing* cycles:							Textual references:
A. Conquest cycle: Wood-Earth-Water-Fire-Metal (e.g. Wood conquers Earth)							
Metal	Wood	Earth	Water	Fire	Metal		*SHD Rishu* A, 83-87b3
Metal *10.2.6.*	Wood *8.12.4.*	Earth	Water *5.9.1.*	Fire *11.3.7.*	Metal		*SHD Rishu* B, 83-87b2
	Wood *A B*	Earth *E F*	Water *I J*	Fire *C D*	Metal *G H*		*SHD Rishu* B, 79-82b2
	Wood	Earth	Water	Fire	Metal	Wood	*Huainanzi* 4:11a
Reversed conquest: Earth-Wood-Metal-Fire-Water (e.g. Earth is conquered by Wood)							
  (Kings)	Earth Yellow Huangdi	Wood Green Yu	Metal White Tang	Fire Red Wen	Water Black	Earth	*Lushichunqiu* 13
	Earth	Wood	Metal	Fire	Water		Mawangdui, "Xingde"
B. Generation cycle 1: Wood-Fire-Earth-Metal-Water							
Earth	Wood *12.4.8.*	Fire *7.3.11.*	Earth	Metal *6.10.2.*	Water *9.1.5.*		*FMT Rishu*
	Wood *12.4.8.*	Fire *3.7.11.*	Earth *7.3.11.*	Metal *6.10.2.*	Water *9.1.5.*		*Huainanzi* 3: 26b
 Water	*A B* Wood	*C D* Fire	*E F* Earth	*G H* Metal	*I J* Water		*SHD Rishu* A, 68-77a2; *Lushichunqiu*; *Huainanzi* 3: 28b; *Huainanzi* 5
	8	7	5	9	6		*Lushichunqiu*; *Huainanzi* 5
	A 1. Wood	*C 1.* Fire	*E 1.* Earth	*G 1.* Metal	*I 1.* Water		*Guanzi* 41; *Huainanzi* 3: 16b
	Green	Red	Yellow	White	Black		*Huainanzi* 3: 16b

C. Generation cycle 2: Wood-Fire-Metal-Water-Earth

	Wood	Fire	Metal	Water	Earth		*Zuozhuan*, Chao 29
	Wood East	Fire South	Metal West	Water North	Earth Center		*SHD Rishu* A, 88-92b3
Earth Yellow	Wood Green	Fire Red	Metal White	Water Black	Earth		*Huainanzi* 4: 11b, 17a-18b

D. Magic square order: Water-Fire-Wood-Metal-Earth / Water-Fire-Metal-Wood-Earth

	Water 1	Fire 2	Wood 3	Metal 4	Earth 5		*Hongfan*
 (months)	North 7,8,9 Red	South 1,2,3 Black	East 10,11,12 White	West 4,5,6 Green			*SHD Rishu* A, 95-98a3
	Water blood	Fire *qi*	Metal sinews	Wood bone	Earth skin	Stone hair	Mawangdui, "Shiwen"

* Only the major correlations in each text are listed, including *Wuxing*, seasons, directions, colors, and day-signs, in their original order. The ten-stem day-signs are represented by capitalized and italicized English letters -- *A B C* for *jia, yi, bing*, etc.- - and the twelve-branch day-signs are represented by numbers, in italics, followed by a period -- *1. 2. 3.* for *zi, chou, yin*, etc. The same system of representing day-signs is used also in Table 3.2.

Redefining the Order of Time

Both conquest and generation cycles, and other less prominent cycles, were articulated to redefine the cosmic order of time and space. The early applications of conquest and generation cycles were most often to the archaic structure of time – day-signs of ten stems and twelve branches. In the case from *Zuozhuan*, in the seventeenth year of Duke Zhao (523 B.C.), quoted earlier in this chapter, the minister of Lu, Shen Xu, predicted that the fire would start not on the *bingzi* day, but on the *yinwu* day.[51] This was because the days were classified into day-signs of ten stems and twelve branches that were correlated with Water, Fire, and so on. As demonstrated in Chapter 2, the structure of time, including the day-sign system, functioned in the earlier cosmology to classify ancestors and to structure rituals of ancestor worship during the Shang. Once incorporated into the cycles of *Wuxing*, the order of time was divorced from the ancestral cult of any particular clan; it became instead a manifestation of the universal cosmic pattern.

That redefining the structure of time was one of the primary functions of *Wuxing* cycles is fully exhibited in the four almanacs (*rishu*) introduced earlier, which provide the earliest evidence with reliable dating of complete conquest and generation cycles of *Wuxing* (see Table 3.1). Through correlation with day-sign systems, the most popular conquest cycle (Wood-Earth-Water-Fire-Metal) and generation cycle (Wood-Fire-Earth-Metal-Water) were completely represented. In the almanac *Rishu* A from Shuihudi (abbreviated SHD.*Rishu* A), a conquest cycle is used to correlate the ten-stem and twelve-branch day-signs:[52]

[*Jia* 甲 and *yi* 乙 days are Wood, Wood conquers Earth.]
Bing 丙 and *ding* 丁 days are Fire, Fire conquers Metal.
Wu 戊 and *ji* 己 days are Earth, Earth conquers Water.
Geng 庚 and *xin* 辛 days are Metal, Metal conquers Wood.
Yin 寅 and *gui* 癸 days are Water, Water conquers Fire.
[*You* 酉],*chou* 丑, and *si* 巳 days are Metal, Metal conquers Wood.
Wei 未, *hai* 亥, [and *mao* 卯 days are Wood, Wood] conquers Earth.
Chen 辰, *shen* 申, and *zi* 子 days are Water, Water conquers Fire.

In the almanacs from Fangmatan 放馬灘 (abbreviated FMT.*Rishu*) the generation cycle is used to correlate the twelve branches:[53]

51 *Zuozhuan*, Zhao 17. See the discussion and translation of this case in the first section of this chapter.
52 Shuihudi Qimu zhujian zhengli xiaozu, 1990: "Rishu B," strips 79b–87b; transcription and commentaries on p. 239.
53 He Shuangquan, 1989.

Wood generates *hai*, the husband of *mao* is *wei*.
Fire generates *yin*, the husband of *wu* is *xu*.
Metal generates *si*, the husband of *you* is *chou*.
Water generates *shen*, the husband of *zi* is *chen*.
Earth generates Wood, Wood does Fire, Fire does Earth, Earth does Metal.

These applications of conquest and generation cycles to the ancient structure of time, as demonstrated in sequences A and B in Table 3.1, established the two basic cycles of *Wuxing* interactions, the most prominent cycles in *Wuxing* theories throughout the later imperial dynasties.

Correlating *Wuxing* with the archaic day-sign system did not simply create a new technique of divination to replace the old one, but rather transformed the structure of time of the ancient cosmology. For the Shang and early Western Zhou ruling clans, the day-sign system was a structure of classification used in ancestor worship, representing a stable ritual order by means of cyclic repetition through time. Such a stable order of the ancestral cult was the foundation of the rulers' legitimacy. The redefining of the archaic day-signs in the *Wuxing* system divorced the structure of time from the ancestral cult, transforming it from a repetitive stability of the ancestral rites of a specific clan to dynamic interactions of cosmic energies constantly conquering and generating one another in a universal pattern.

Redefining Spatial Order

Besides transforming the archaic structure of time, the other primary application of the new *Wuxing* system was to redefine the spatial structure of *Sifang*. As demonstrated in Chapter 2, the spatial structure of *Sifang* defined an absolute religious and political supremacy of the center, where divine knowledge acquired through ancestral cults and political dominance were reinforced by the alien status and subordination of the *Sifang*. The *Wuxing* system incorporated *Sifang*, redefining *Sifang* and center as five equal spatial units, and thus totally rejected the supremacy of the center.

Such a transformation of the *Sifang* spatial structure produced some of the other sequences of *Wuxing*, the influence of which can be observed right down to the Qin and Han Dynasties. One sequence so produced was the second generation cycle in Table 3.1 (Wood-Fire-Metal-Water-Earth). Almanacs from Shuihudi (SHD.*Rishu* A) clarify that this order came from imposing *Wuxing* on the four *fang* in a clockwise sequence and adding the center as the fifth at the end:[54]

> Eastern *fang* is Wood, Southern *fang* is Fire, Western *fang* is Metal, the Northern *fang* is Water, the Center is Earth.

Though not named as a generation cycle in SHD.*Rishu* A, this sequence was clearly identical to a generation cycle in an early Han text, *Huainanzi*, that represents a much more systematic *Wuxing* system:[55]

> Earth when refined [subjected to change] produces Wood, Wood when refined produces Fire, Fire when refined produces clouds [of metallic *qi*], clouds when refined produce Water, and Water when refined reverts to Earth.

That the second generation cycle is derived from imposing *Wuxing* on a *Sifang*-center cosmology provides a clue to the same sequence of *Wuxing* recorded in *Zuozhuan*, from the twenty-ninth year of Duke Zhao, which ordered the offices of *Wuxing* into the sequence of Wood, Fire, Metal, Water, and Earth. This sequence found in *Zuozhuan* could be the earliest representation of this space-oriented generation cycle.

Imposing *Wuxing* on the *Sifang*-center structure produced yet another sequence, Water-Fire-Wood-Metal-Earth (sequence D in Table 3.1). Although not widely adopted later, this sequence connects *Wuxing* to the magic squares and their numerology. Graham has persuasively illustrated how magic squares (*Hetu* 河圖 and *Luoshu* 洛書) and the numerology of *Wuxing* were, in all likelihood, derived from correlating *Wuxing* and numbers with the center and the four directions.[56] To summarize Graham's structural analysis, the sequence of Water-Fire-Wood-Metal-Earth resulted from counting the four directions from the throne in the North from which the ruler faces South, with the central position the one left over from binary divisions such as East/West and North/South and serving as the position from which the oppositions are drawn:[57]

South2

East3 Center5 West4

North1

54 Shuihudi Qimu zhujian zhengli xiaozu, 1990: "Rishu A," strips 83b3–92b3; transcription and commentaries on p. 223.

55 *Huainanzi* 4, p. 11b; references are to Liu Wendian, *HNHLJJ*; translation adapted from Major, 1993, p. 186.

56 Graham, 1989, pp. 342–49.

57 I adopt the following diagrams from Graham, ibid., with minor alterations.

Imposing *Wuxing* on this spatial and numerical system led to the numerical correlation of *Wuxing* represented in "Hongfan."[58] The numbers following this correlation continued into a second cycle resulting in the following numerical system of *Wuxing*:

	2, 7 Fire	
3, 8 Wood	Earth 5	Metal 4, 9
	Water 1, 6	

With further elaboration, this diagram can be identified as the magic square called the "river chart" (*Hetu*) dating from the Song Dynasty (960–1279 A.D.).[59] Reading from Earth at the center through Metal and proceeding clockwise gives the first generation cycle (Earth-Metal-Water-Wood-Fire-Earth). Reversing Fire and Metal and reading through Wood results in the conquest cycle in its passive form, Earth-Wood-Metal-Fire-Water:

	4, 9 Metal	
3, 8 Wood	Earth 5	Fire 2, 7
	Water 1, 6	

Extending this diagram from the cardinal to the intermediate points and filling those points with even numbers forms the diagram of *Luoshu*, the magic square with numbers adding up to 15 in every direction:

58 The date of "Hongfan" is extremely controversial. According to the recent monograph on this text by Michael Nylan, the original "Hongfan" was a ruler's manual on administration during the Warring States period, dated to the late fourth century B.C. See Nylan, 1992, pp. 13–44, 105–48.

59 For *Hetu*, *Luoshu*, and the magic square, see also Needham and Wang Ling, 1959, vol. 3, pp. 56–8; Major, 1984.

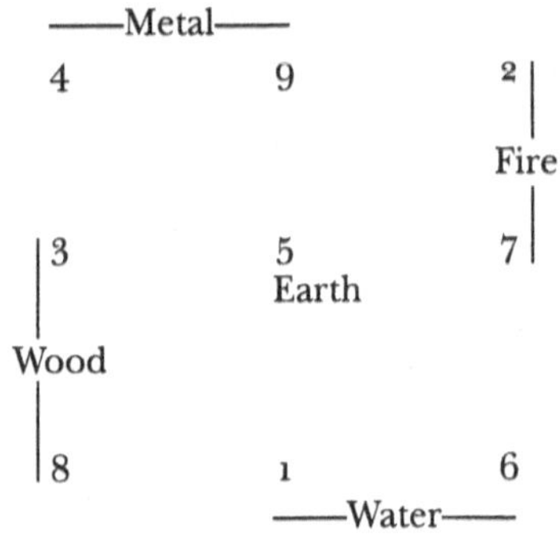

Graham imagines the tremendous impression made by the discovery of this mathematical secret of the cosmos, attributing the discovery to the numerologists' search for "some fascinating symmetry."[60] The intellectual process of discovery revealed by Graham's structural analysis is fascinating in itself; and its significance in transforming cosmology, and the sociopolitical implications of such a transformation, are equally revolutionary.

Redefining the spatial concept of *Sifang*-center cosmology and the day-sign system, *Wuxing* provided a new conception of time and space. The four directions (*fang*) and day-signs in the Shang and Zhou context functioned as spatial and temporal structures for rituals centered on ancestor worship. *Wuxing* introduced to these structures the dynamic movement of cosmic energies; interaction and change replaced centrality and hierarchy as the fundamental properties of time and space. The day-sign system used in Shang and Zhou rituals represented the stability and eternity of ancestor worship. Once correlated with *Wuxing*, it became the rhythm of the dynamic cosmic movement that brought along changes in daily life. The spatial structure underwent the same kind of change. The four *fang* and the center had been a static hierarchy, with the center being the "zone of the sacred."[61] Now redefined as five spatial forms of interactive cosmic energy, *Sifang* and the center became five equal units that existed in ceaseless interaction with one another. In this transformation, the sense of eternity and permanence in the cosmology were not lost but simply transformed. What remained eternal and permanent was no longer a cosmological and political center, but change itself and the patterns of change.

The center in this sense has lost its superiority and sacredness. Center/Earth can be conquered by East/Wood, just as West/Metal is

60 Graham, 1989, pp. 348–9.
61 The term is borrowed from Marcel Eliade's theory of symbolism of the center. See Eliade, 1955.

conquered by South/Fire. The day-signs associated with Center/Earth – *Wuji* days – are simply two of ten stems and have no superiority over the other eight. In Table 3.1 we see that in pre-Han mantic texts and some Han texts, Earth was not necessarily located at the initial or central position of the sequence. It took great effort by theorists during the imperial period to restore the superiority of Center/Earth. Yet it is worth noticing that the superiority of Center/Earth was not initially a feature of *Wuxing* when first introduced in social practices. Quite the contrary, *Wuxing* was articulated to undermine the centrality of the center in the *Sifang* cosmology. The restoration of the centrality of Earth was a later development, part of the ideology of the centralized empire of Han – a subject that shall be addressed in Chapter 4.

Correlations: New Connections between Heaven and Man

While *Wuxing* contributed to the destruction of the old center of both cosmology and political structure, it was also indispensable for constructing the new form of political power. It created a new concept of human sovereignty and political order by building an all-encompassing cosmology that directly correlated the realm of Heaven with the realm of Man, facilitating the change in the conjunction of political power and divine authority.

In transforming *Sifang* cosmology, the political actors of the time used *Wuxing* to deny the hereditary king's monopoly on communication between Heaven and Man through the ancestral cult. At the same time, they used *Wuxing* to create means of communication with Heaven that directly correlated every aspect of human life with the cycles of cosmic movement, along with omen reading based on those correlations. In such a correlative system, messages between the divine and human beings could be passed directly, without the need for royal ancestors as mediators. Such direct correlation between Heaven and Man redefined rulership, and eventually created a new concept of human sovereignty for the territorial states and the succeeding empire.

Inventing Rulership

While using cosmological discourse to deny the hereditary king's divine authority, various social-political groups – court religious specialists, ministers, bureaucratic officials, and scholars – were also using cosmology as the ultimate means of creating a new concept of rulership. While the philosophers were debating the nature of an ideal ruler in terms of the "sage" king, using a "bare cosmological scheme"[62] as seen in the *Analects* 論語, *Tao De Jing* 道德經, or *Mencius* 孟子, other political actors in various territorial courts were already using correlative schemes to invent new concepts of rulership, as found in *Zuozhuan.*

Physicians, musicians, cooks, and ministers in the court used correlative cosmology to separate the king's body from direct connection with the spirits of his ancestors, placing it under the regulation of cosmic patterns accessible through their own professional knowledge. This is best illustrated by the interpretation of an illness of the duke of Jin. The duke of Jin was ill, and minister Zi Chan of Zheng was invited to Jin to be consulted. The duke of Jin explained that he had consulted the oracle diviner, who said that the illness was brought by two spirits, Shi Chen 實沈 and Tai Tai 臺駘; but the scribes no longer knew what these spirits were. Zi Chan first recounted the legend that the two spirits were the spirits of asterisms Chen 辰 (Great Fire, Sco α) and Shen 參 (Orion), which the ancestors of Shang and Tang had followed in order to prosper. But after recounting this ancient legend about the spirits and ancestors, Zi Chan stated that these spirits could only bring harm in the realm of nature, such as drought, flood, and unseasonable rain or snow, but could not possibly harm "the ruler's body." This was because the ruler's body was affected by the compliance of his bodily actions – such as travel, food, and emotions – with the cosmic order, which had nothing to do with the spirits. He hence described the cosmic order for a ruler in terms of regulating time and reserving the cosmic energy – *qi* 氣:[63]

> Gentleman [divides the day] into four periods – the morning, to hear the affairs of the government; noon, to make full inquiries about them; the evening, to consider well and complete the orders [he has resolved to issue]; and the night, for rest. In doing so, he regulates and dissipates the *qi* of his body, so that it is not allowed to get shut up, stopped, and congested, so as to injure the body.

Hearing such a cosmological interpretation of his own illness, the duke of Jin praised Zi Chan as "a gentleman with profound knowledge" and rewarded him handsomely. A more sophisticated cosmological theory about the ruler's body was fabricated by the physicians. For the same illness of the duke of Jin, physician He from Qin was invited for consultation. The physician diagnosed the duke's illness as resulting not from spirits or food, but from his closeness to women. He explained this according to the cosmological theory of the six *qi*, five tastes, five colors, four seasons, five regulations, and *Yin-Yang*:[64]

62 Graham, 1986, p. 91. 63 *Zuozhuan*, Zhao 1. 64 Ibid.

Heaven has six *qi*, which descend and produce the five tastes, go forth in the five colors, and are verified in the five notes; but when they are in excess, they produce the six diseases.

The six *qi* are the *Yin*, the *Yang*, wind, rain, obscurity, and brightness. In their division, they form the four seasons; in their order, they form the five regulations. When any of them is in excess, there ensues calamity . . . Now your lordship pays no regard to such regulations or to time, how could you possibly not be sick.

Physician He was rewarded generously and praised as "a fine physician" for his diagnosis by a minister of the state.

According to both the "gentleman of profound knowledge" Zi Chan and the "fine physician" He, the king's body was no longer the mediator between the ancestral spirits and the world of the living, but an expression of a cosmic pattern that regulated both natural and social orders. Unlike the hereditary king, whose authority lay in his direct connection to his ancestral spirits, the new territorial ruler was to imitate the universal pattern of the cosmos in his actions, by "regulating" the cosmic energy in his body and by following the order of time in his actions.

Not only the king's body was regulated by the cosmic order, but also his political power and his position in the power structure. At a time when the political power of hereditary lords of the former Zhou was constantly being usurped by the rising ministers and military leaders, the dynamic cosmic movement provided the best rationale. Duke Zhao of Lu was driven away from his state by minister Ji, and died in exile. A minister of Jin, Zhao Jianzi, was puzzled by the fact that the Lu people as well as lords of other states did not accuse Ji of usurping the power of his lord, but instead supported him. In answering Zhao Jianzi's question, scribe Mo of Jin used "the way of Heaven" to explain the shifting of power away from the noble lords to the ministers:[65]

Things are born in twos, in threes, in fives – in pairs or double. Hence in Heaven there are the three *Chen* 辰, in Earth there are five *xing*; the body has the left [side] and the right, and every one has his mate. Kings have their dukes, and princes have their ministers who are their doubles. Heaven produced the Ji family to be the double of the marquis of Lu, as has been the case for long . . . A state does not maintain the same altar forever, rulers and ministers do not remain in permanent positions; from the ancient times it has been so.

According to this cosmological explanation, the power of the hereditary ruler, which formerly had been considered sacred and eternal, now became just another manifestation of fluid cosmic movement. Like the

65 *Zuozhuan*, Zhao 32.

movement of the sun, moon, and stars in Heaven, and like the cyclic conquest and generation of the *Wuxing* on earth, the position of ruler-subject is never permanent, but subject to constant change.

While the cosmological discourse redefined the king's body and his political position, it also reinterpreted the primary function of the ruler and the state, that is, the performance of rites. For the Shang and Zhou rulers, the two "great affairs" of the state were ritual sacrifice and warfare; both functioned to offer flesh and blood in ceremonies for the ancestral spirits to consume. Such monopolized sacrificial rites to ancestors were the source of the ruler's political power. To define the nature of the new territorial ruler and territorial states, a new theory of rites was needed, identifying ritual as the ruler's primary function. Correlative cosmology became indispensable to such theorizing. A minister of Zheng elaborated ex-minister Zi Chan's account of rites into a complex cosmological theory:[66]

> I have heard our late great minister Zi Chan say, "Rites are the patterns of Heaven, the righteousness of Earth, and the actions of people." Heaven and Earth have their patterns, and people take these for their pattern, imitating the brilliance of Heaven, and according with the nature of the Earth. [Heaven and Earth] have produced the six *qi*, and make use of the five *xing*. The *qi* became the five tastes, manifested in five colors, and displayed in the five notes. When these are in excess, there ensue obscurity and confusion, and people lose their nature. That is why there were rites made to support that nature . . . There were ruler and minister, high and low, in imitation of the righteousness of Earth. There were husband and wife, with the home and the world outside as spheres of their divided duties. There were father and son, elder and younger brother, aunt and sister, maternal uncles and aunts, father-in-law and connections of one's children with other members of their mother's family, and brother-in-law – all to resemble the brilliance of Heaven. There were affairs of government, the control of labor, administration and services – in accordance with the four seasons. There were harsh punishment and fearful prisons, making the people stand in awe, resembling [Heaven's] violent forces of thunder and lightning. There were gentleness, kindness, generosity, and harmony, in imitation of the producing and nourishing action of Heaven.

Reinterpreted through such cosmological discourse, ritual was no longer a matter of feeding the ancestral spirits; it acquired instead the function of maintaining order and hierarchical relations in the human world, and of expressing and regulating human emotions that were part of the cosmic *qi*. The role of the ruler was to observe the ritual rules as his way of government, in order to imitate the pattern of Heaven and Earth and to "support the nature of Heaven and Earth."

66 *Zuozhuan*, Zhao 25.

Texts and Correlative System-Building

To deprive the ruler of his direct connection with the spiritual world, to disperse the king's divine authority into their own hands, and to subject him – his body, power, and function – to the authority of the cosmos, the religious and natural experts, military professionals, bureaucrats, and scholars all participated in building correlative systems that incorporated every dimension of human existence, correlating them with the natural world and with cosmic patterns.

The construction of such correlative systems paralleled the change in the forms and media employed to conceive and represent cosmology. Shang cosmology was manifested primarily in ritual and political action, in divination, in physical construction of tombs and cities, and in visual images of engravings and the arts. The existing written records representing the cosmology are divination records and bronze inscriptions meant to communicate with ancestral spirits. Even in the Western Zhou, when long inscriptions become available, the cosmology is still reflected through its ritual context, since the inscriptions were cast on ritual vessels as messages to ancestral spirits.[67] By contrast, in the Warring States period, texts increasingly become the authoritative source of cosmologies that served as models for the new political order and ritual rules. What is more essential, the textualized cosmology constitutes the discourse of political ethics and political criticism, which is not the case in Shang and Western Zhou material.

This change reminds us of the dichotomies in modern Western theories, such as that of "concrete" versus "abstract" modes of thought proposed by Lévi-Strauss, and that of "oral" versus "literary" modes of communication employed by Jack Goody.[68] John Henderson and Sarah Allan both adopt Goody's theory, attributing the development of *Wuxing* out of the Shang cosmology of *Sifang* to the growth of literacy.[69] Allan distinguishes Shang thought as being "mystic" or "implicit" – not subject to the self-awareness and conscious analysis inevitable with the growth of a literary tradition – while the later patterns of thought were "explicit," resulting from a literary tradition. I choose not to apply these models, for both factual and theoretical reasons. Factually, both concrete and abstract modes of thinking existed in the Shang as well as in the Warring States period, and literacy was not a Warring States invention – it was highly developed in the Shang and Western Zhou in the form of oracle and bronze inscriptions, and possibly also as *ce* 冊 and

67 Falkenhausen, 1993, pp. 146–52.
68 Lévi-Strauss, (1962) 1966; Goody, (1977) 1987.
69 Henderson, 1984, pp. 1–46; Allan, 1991, pp. 13–14, 171–6.

dian 典. Theoretically, the fact that literacy became the form of abstraction, of conscious analysis and criticism, was not the "cause" of the transformation of cosmology, as Jack Goody would argue; rather, the change in the function of literacy was itself a part of the total historical change in which cosmology and social reality as well as the mode of representation were transformed. What I propose is an alternative model of interpretation, referring to the *Sifang* cosmology of the Bronze Age as a system of ritual and political action, while the *Wuxing* cosmology of the formative age of imperial China was a text-based political discourse.

That cosmology became a text-based discourse for diverse political groups coincided not only with the growing number of texts circulated among various professions and groups, but also with the changing nature of literacy. All the rising political groups we have discussed relied on a body of texts that carried the authority and tradition of their professions. For example, religious and natural experts developed a body of texts that incorporated various menus, charts, diagrams, catalogues, calendars, as well as theories that were used in their professions. Military professionals created a large body of military treatises. The appearance of these treatises themselves, according to Lewis, suggests a radical change in power relations, since "the military specialists consciously viewed their 'textual' vision of warfare as a denial of the chivalric warfare of the old nobility."[70] Scholars of philosophy especially placed the greatest authority on the body of texts that identified their own philosophical traditions, such as the Confucian texts of the "six arts" – *Yi* 易, the *Book of Poetry* 詩, the *Book of Documents* 書, *Rites* 禮, the *Spring and Autumn*, and the *Music* 樂. The bureaucrats, too, relied on texts for standardizing bureaucratic theories, rules, and procedures, and used them to guide their daily performance in office. Li Ling sees such growth of texts as a result of the dissemination of the official learning of the court downward into private learning and popular practice during the Warring States period, a phenomenon that Karl Jaspers called "the breakthrough of the Axial Age."[71] Such a phenomenon is not simply the "growth of literary tradition," but rather a change in the nature of literacy, from a medium passing messages between the living king and spirits to a medium that formed a text-based discourse circulating among diverse groups in society.

Because the construction of correlative systems first appeared in such diverse bodies of texts belonging to different professional and intellec-

70 Lewis, 1990, p. 98.
71 Li Ling, 1993, pp. 2–10.

tual traditions, they involved most major classifications used in different professions. Through the Warring States period and into the early Han, multiple correlative systems coexisted, competed, and even contradicted one another. Each correlative system, too, remained incoherent and unsystematic in different textual versions. In this process, the old structures of time and space, the symbolic systems such as four directions, four seasons, twelve months, color symbolism, and so forth, changed their meanings and functions and became important correlative schemes.

That the old classification and symbolic systems were transformed to create correlations in this system-building process can be illustrated by the silk manuscript excavated in Changsha in southern China, the earliest calendrical divination manuscript of the Warring States period that survives[72] (Fig. 3.1). The text itself is arranged according to the *Sifang* structure. At the four sides of the manuscript was a calendar of the twelve months, accompanied by the images and names of monthly gods along with monthly prohibitions, all of which were divided into four seasons and arranged in four cardinal directions. The center is occupied by two parts, written in opposite directions. The part that is longer (thirteen lines) warns repeatedly against disruption of the pattern of time and lists all kinds of calamities that such disruption would induce. The shorter part (eight lines) accounts for the myth of the formation of this pattern of time. This manuscript does not have the terms *Wuxing*, Fire, Water, and so on, but it does mention five trees (*wu mu* 五木), each having the symbolic color associated with *Wuxing*, five evil omens from nature (*wu yao* 五妖), and five officials (*wu zheng* 五政), who are likely the same five officials in charge of *Wuxing* found in *Zuozhuan*, Zhao 29.

More important than the appearance of the categories of five, the *Sifang* structure itself has undergone fundamental changes. Even though the manuscript is in the shape of the old *Sifang* structure, this structure differs from that found in Shang cosmology. For example, *Sifang* in Shang oracle inscriptions was primarily a spatial concept from which a structure of time was derived. In this manuscript, the structure of the four seasons takes primacy over the spatial structure of *Sifang*, while keeping its shape. The concern in three parts of the text is with the pattern of time; the spatial structure becomes secondary. This shifting priority results in substituting the dynamic movement of time for a static

72 This study of *Chu boshu* is based primarily on Li Ling's transcription. See Li Ling, 1985. Other major transcriptions and studies of this text were also consulted, including Rao Zongyi and Zeng Xiantong, 1985; Noel Barnard, 1973.

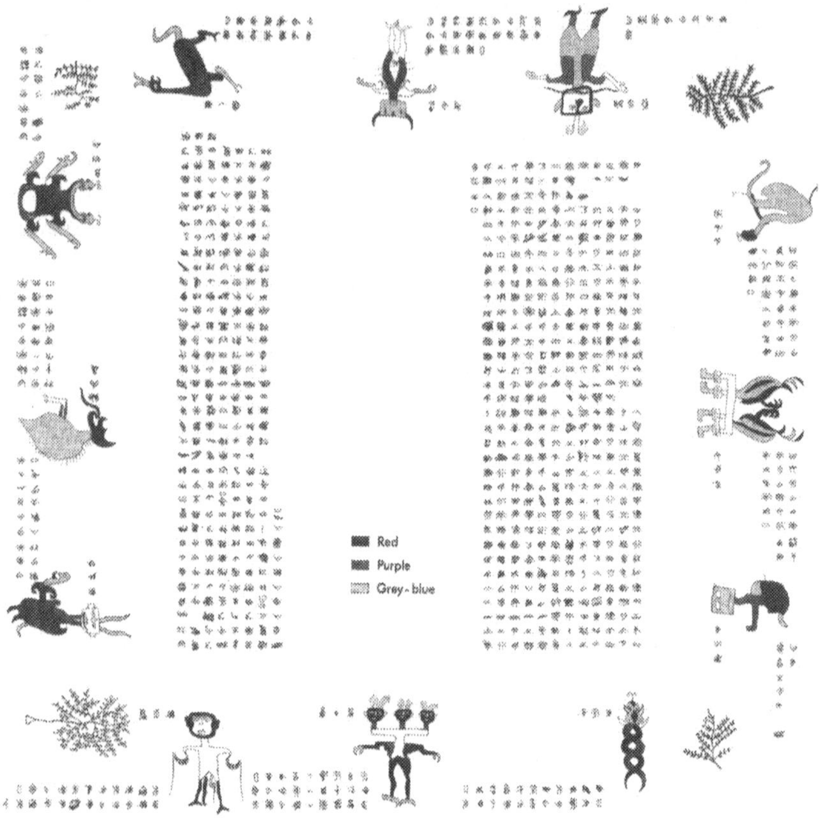

Figure 3.1. The reconstruction of *Chu boshu* (Barnard, *The Chu Silk Manuscript*, Canberra: The Australian National University, 1973).

structure of space, and in eliminating the notion of an eternal fixed center, since time itself is fluid.[73]

The function of *Sifang* has also changed in this calendrical manuscript. While the *Sifang*-center structure had enabled the Shang king to monopolize ritual communication with the high god through the royal ancestral line, the same structure in this calendrical manuscript served to build correlations between human activities and the cosmos, making mediation by royal ancestors unnecessary. The four directions became

73 Various mantic practices that flourished in the Warring States and early Han, such as "*Xingde*" and "*Kanyu*," reflect this preoccupation with time. For a discussion of "*Xingde*," see Major, 1987. For the study of "*Kanyu*" from its earliest records to its relation to *fengshui* of later times, see Loewe, 1983–5.

the basis for correlating major categories such as seasons, months, colors, gods (*di*), activities, prohibitions, and omens induced by activities that violated the cosmic order. Such correlation facilitated the local lords in their competition for power. The activities listed in the calendar, and monthly prohibitions for the lords, exhibited this new function. Such activities included military actions (six references), meeting with other lords for allies (one reference), constructing cities and buildings (three references), taking in and marrying off women (three references), ritual ceremonies (two references), and the execution of wicked people (two references). The success of these activities was directly correlated to the cosmic pattern of time. If carried out at the wrong time, they would induce catastrophes and disorder in the universe, including landslides, floods, disorders in the moving patterns of heavenly bodies, abnormal growth of vegetation, unseasonable rain and storms, robbery, and calamities and chaos in the state. These catastrophes and disorders became omens – signs signifying a disruption of the divine cosmic order by human activities.

Sifang in this Chu calendrical manuscript is only one of many symbolic systems transformed in constructing the Heaven-Man correlation. In such constructions, concepts such as directions, seasons, day-signs, and colors changed their functions and meanings while retaining their antique forms. Sharing with *Wuxing* the same functions, they stood for cosmic movements and correlated human actions with these movements.

That the cosmic cycles and correlations of the *Wuxing* system were represented by multiple classification systems is demonstrated repeatedly in a series of texts dated between the late Warring States period and the early Han. These texts include military treatises such as *Sun Bin bingfa* and *Sunzi bingfa* excavated from Yinqueshan 銀雀山 (abbreviated YQS *Sunzi bingfa*), mantic manuscripts such as the Chu calendrical manuscript (*Chu boshu* 楚帛書) and the almanacs from Shuihudi (abbreviated *SHD.Rishu*), and philosophic texts such as *Mozi*, *Guanzi* 管子, and *Huainanzi*. Table 3.2 shows that in these texts, the major *Wuxing* cycles and correlations were often represented by alternative symbolic systems – day-signs, colors, seasons, and directions – without the presence of *Wuxing* terminology. Textual references in Tables 3.1 and 3.2 reveal that the same text, or texts of the same period, often used *Wuxing* and other alternative symbolic systems to represent these cycles and correlation.

Day-signs were one such alternative system. A passage in SHD.*Rishu* A on sickness (first reference under generation cycle 1, Table 3.2) is based on ten-stem day-signs. Without the presence of the terms Wood, Fire,

Table 3.2. Wuxing *cycles and correlations represented by multiple symbolic systems*

Sequence of *Wuxing* cycles:						Textual References:
A. Conquest cycle: Wood-Earth-Water-Fire-Metal						
	Green	Yellow	Black	Red	White	*Sun Bin bingfa*
B. Generation cycle 1: Wood-Fire-Earth-Metal-Water						
	A B	*C D*	*E F*	*G H*	*I J*	*SHD Rishu* A, 68-77a2
	East	South	Center	West	North	
	Green	Red	Yellow	White	Black	
(cause of sickness)	meat	chicken	fish	dog	dry meat	
	A B	C D	------	E F	G H	*Mozi*
	East	South	------	West	North	
	Green	Red	------	White	Black	
	Green	Red	Yellow	White	Black	*Chu boshu*
	Spring	Summer	------	Autumn	Winter	
	East	South	------	West	North	
	East	South	Center	West	North	*Guanzi* 8
	Spring	Summer	------	Autumn	Winter	
	Green	Red	Yellow	White	Black	
	8	7	5	9	6	
(*qi*)	dry	*Yang*	harmony	wet	*Yin*	
	East	South	Center	West	North	*Guanzi* 40
	Spring	Summer	------	Autumn	Winter	
(*qi* produces)	Wind	*Yang*	------	*Yin*	Cold	
	Wood	Fire	Earth	Metal	Water	
	bone	*qi*	skin	nails	blood	
	East	South	Center	West	North	*Huainanzi* 3: 5b-6b; *Huainanzi* 5
	Wood	Fire	Earth	Metal	Water	
	Spring	Summer	*Sifang* [mid-summer]	Autumn	Winter	
C. Generation cycle 2: Wood-Fire-Metal-Water-Earth						
(gods)	Yellow	White West	Black North	Green East	Red South	*YQS Sunzi bingfa* (in reversed order)
	East	South	West	North	Center	*Huainanzi* 3: 9a-10a
(officials)	agriculture	military	public order	public works	metropolitan affairs	
D. Magic square order: Water-Fire-Wood-Metal-Earth						
(notes)	*yu*	*zhi*	*jue*	*shang*	*gong*	*Huainanzi* 3: 21b (in reversed order)
	Water	Fire	Wood	Metal	Earth	

Water, and so on, these day-signs stood for *Wuxing*, representing both conquest and generation cycles and correlations with colors and directions. The order of the day-signs was in the generation cycle *jia yi* (Wood) – *bing ding* (Fire) – *wu ji* (Earth) – *geng xin* (Metal) – *yin guei* (Water). The causes and cures of sickness, nevertheless, were based on a conquest cycle. If one got sick on the *jia* and *yi* days (Wood), the sickness was from the Eastern *fang*, from eating meat correlated with the East, and one would die in the color green. On *wu ji* days (Earth) one would get worse, because Wood conquers Earth. To be cured, one would have to rest and perform the *zuo* ritual on *geng xin* days (Metal), because Metal conquers Wood.

The same system is found in *Mozi* (the second reference under generation cycle 1, Table 3.2), a philosophical text that incorporated a fragment of calendrical divination:[74]

On the *jia yi* days the god kills the green dragon in the Eastern *fang*.
On *bing ding* day he kills the red dragon in the Southern *fang*.
On *geng xing* day the white dragon in the Western *fang*.
On *yin gui* day the black dragon in the Northern *fang*.

This system is almost identical to the *Rishu* passage quoted earlier, except for the omission of *wu ji* days/center/yellow, which reflects the earlier solution of merging ten heavenly stem day-signs with the four directions.

Like the day-signs, colors represented the cycles and correlations of the *Wuxing* system. Two military texts unearthed at Yinqueshan and dated to the mid Warring States period can best illustrate this color symbolism. One of these texts, *Sun Bin bingfa* (the reference under conquest cycle, Table 3.2), has a conquest cycle of five colors of soil, which is identical to the conquest cycle of *Wuxing*: "Green conquers Yellow, Yellow does Black, Black does Red, Red does White, White does Green."[75] The other military text, *Sunzi bingfa* 孫子兵法 (the first reference under conquest cycle, Table 3.2), mentions five *Di* (gods) symbolized by five colors and directions:[76]

Yellow *Di* went
to the South to attack Red *Di* . . .
to the East to attack Green *Di* . . .
to the North to attack Black *Di* . . .
to the West to attack White *Di* . . .
And after having conquered the four *Di* he possessed all under Heaven.

74 *Mozi*, ch. 47. 75 Yinqueshan Hanmu zhujian zhengli xiaozu, 1975a, p. 61.
76 Yinqueshan Hanmu zhujian zhengli xiaozu, 1975b, p. 101. For bibliographical information on this text, see Gawlikowski and Loewe, 1993.

Tracing Yellow *Di's* expedition back, one gets Earth/Yellow – Metal/White – Water/Black – Wood/Green – Fire/Red, which happens to be the reversed generation cycle 2 of Table 3.2. It also happens that the order of five colors of trees in *Chu boshu* is identical to that of generation cycle 1 (the third reference under generation cycle 1).

The directions, above all, most often represent correlations and cycles that are identical to those of *Wuxing*. The calendars incorporated in Chapters 8 and 40 of *Guanzi*[77] and in Chapters 3 and 5 of *Huainanzi*[78] all have the directions, *Sifang* plus center, as the primary correlation representing generation cycle 1 (the last three references under generation cycle 1, Table 3.2). In *Guanzi* 8, we found most of the major correlations, such as numbers, colors, seasons, directions, and so on, except for *Wuxing* terminology, all structured on four directions and the center (Fig. 3.2). *Guanzi* 40 was also based on four directions plus the center, with *Wuxing* terms only as subsidiaries generated by the five *qi* of directions and seasons.[79] In *Huainanzi* 3 and 5, the five directions and five *xing* finally merge to be the prime correlation of all other fives.

The frequent repetition of such coherence between old symbolic structures and the new *Wuxing* system can be explained, as Graham has suggested, by a linear chronology according to which these correlations were first built on the archaic systems of seasons and directions, on which *Wuxing* was later imposed.[80] But one must go further to ask why, in the first place, these "archaic systems" started the process of building correlations and representing dynamic cycles of cosmic interactions, a process identical to the development of *Wuxing*; and furthermore, why it was necessary to impose *Wuxing* on these archaic systems.

My explanation is that both the emergence of *Wuxing* and the change in old spatial-temporal structures resulted from the same search for new ways of connecting Heaven and Man and the same process of change in the mode of connecting political power and cosmology. When the former power center collapsed and royal ancestor worship declined, people involved in the new power structure – warlords, ministers, military commanders, diviners, astrologers, physicians, bureaucratic

77 The text is composed of political theories of master Guan Zhong and his followers in the state of Qi during the late Warring States period. *Guanzi*, SBBY edition; for translation see Richett, 1985.

78 According to Major's dating, these chapters were completed with *Huainanzi* as a whole and submitted to Emperor Wu in 139 B.C. See Major, 1993. But Loewe suspects that some *Huainanzi* chapters, including Chapters 3, 4, and 5, were additions written as late as 122 B.C., protesting the policy of the central government. This thought is expressed in Loewe's correspondence with the author and in Loewe, 1995.

79 Graham, 1986, pp. 85–6.

80 Ibid, pp. 84–90.

Figure 3.2. The reconstruction of the "Dark Palace," Chapters 8 and 9 of *Guanzi* (Guo Moruo, Wen Yiduo, and Xu Weiyu, eds., *Guanzi jijiao*, Beijing: Kexue chubanshe, 1956).

officials, and philosophers – were all looking for alternative access to the divine world and knowledge about forces and orders beyond human control. Among the many systems and techniques so developed, cycles of cosmic movement and a direct correlation of the human world with the cosmic order became the two most essential common features. These cycles and correlations were used to transform the received cosmology and symbolic systems. If we postulate such a synchronic social phenomenon and its diachronic development rather than the linear evolution of an intellectual idea, the repetitive similarities between the emerging *Wuxing* system and the transformed symbolic systems, and their integration into a new cosmology, become not only possible but inevitable.

Synthesizing Cosmology and Theorizing Sovereignty

If it was the religious and natural experts, like the scribe Mo, diviner Pi Zao, and physician He, who first formulated the correlative cosmologies, and if it was the various rising political groups who collectively transformed cosmology into a text-based political discourse and built multiple correlative systems, it was the scholars – the masters and students of philosophy who compiled texts and became the architects and civil servants of the imperial apparatus that was taking shape – who accomplished the total integration of multiple correlative systems into one coherent and all-encompassing correlative cosmology, in which *Wuxing* became a prime system of synthesis. Furthermore, if it was the various rising political groups in the Warring States period who collectively usurped the divine authority of the hereditary king and subjected the new rulers to the authority of the cosmos, it was the scholars who completed the formulation of a philosophy of human sovereignty, a theoretical preparation for the unified empire. The synthesis of cosmology and formulation of a philosophy of sovereignty formed a single project of the scholars of the late Warring States period and early Han.

The construction of these correlative systems began in diverse genres of texts concurrently, with various aspects of human life incorporated separately into different systems, serving the diverse interests of different groups and professions. The first three columns in Table 3.3 show such concurrent correlation-building. The correlation in *Zuozhuan* seems to have begun with two or three categories at most, like those of Water and Fire, along with the constellations and the states. There are many categories of fives and fours in this text, but not only were the fives and fours not integrated into a uniform system, there was also a lack of obvious correlation among the categories that were grouped into fives. Correlations built on different systems often coexisted in the same text. In *Chu boshu*, *Mozi*, and Yinqueshan *Sunzi bingfa*,[81] major correlations are built on four directions and seasons, including colors, months, day-signs, and monthly prohibitions. Yet the four directions as the basis for such correlations coexist with categories in five, such as five colors, five *Di* (gods), five *zheng* (officials), and so forth. The same parallel existence of sets of four and five as the bases of correlation is also common in the three *Rishu*. As reflected in Table 3.3, the parallel existence of different

81 Yinqueshan Hanmu zhujian zhengli xiaozu, 1975b, p. 101.

Table 3.3. *The construction and integration of correlative systems*

Correlation	*Zuozhuan*	*Mozi* 47 & *YQS Sunzi*	*Chu boshu*	4 *Rishu*	*Guanzi* 8 (& 40.41)	"Yueling" (*HNZ* 5)
Directions	4 *fang*	4 *fang*	4 directions	4 *fang* + center	4 *fang* + center	5 *wei* (positions)
Seasons	4 seasons	4 seasons	4 seasons	---------	4 seasons (+ year/center *GZ* 41)	5 seasons
12 months	grouped in 4 seasons	---------	grouped in 4 directions / seasons	grouped in 4 directions	---------	grouped in 5 *wei* & *de*
12-branch day-signs	in fragments corr. with *xing*	---------	---------	grouped in 4 *xing*	---------	grouped in 5 *wei* & *de*
10-stem day-signs	in fragments corr. with *xing*	in 4 *fang*	---------	grouped in 5 *xing*	---------	grouped in 5 *wei* & *de*
Xing (phases)	5 *xing* & 5 materials	---------	---------	5 phases	5 *xing* (*GZ* 41)	5 *de*
Colors	5 colors	of 4 *fang* (*MZ*) of 5 *di* (*SZ*)	5 colors	5 colors & 4 colors	ruler's 5 ritual colors	5 colors
Musical notes	5 notes	---------	---------	5 notes	ruler's 5 hearings	5 notes
Pitch pipes	6 pitch pipes	---------	---------	fragments	---------	12 pitch pipes corr. with 5 *de*
Tastes	5 tastes	---------	---------	---------	ruler's 5 tastes	5 tastes
Numbers	---------	---------	---------	1,3,5,7,9 corr. with 5 notes	8,7,5,6,9 corr. with 5 *fang*	8,7,5,6,9 corr. w/ 5 *wei* & *de*
Palace	---------	---------	text in the *ya* shape	---------	9-room shape of the text	9-room *Ming tang* used in 12 months
Gods (*di*)	---------	5 *di* (*SZ*) in 5 colors	Yandi	---------	5 queens (Huangdi *GZ* 41)	5 *di* with names (*HNZ* 3)
Spirits	5 officials	4 dragons in 4 colors (*MZ*)	5 officials & Zhu Rong	many spirits & ghosts	6 assistants of 6 *fang* (*GZ* 41)	5 assistants identical to 5 officials
Creatures	---------	---------	---------	---------	5 classes	5 classes
Human body	---------	---------	---------	---------	in 5 parts (*GZ* 40)	in 5 parts
Heavenly bodies	*fenye* system	---------	*fenye* system	28 *xu* corr. with 12 months	stars, sun, year star, *chen*, moon	28 *xu* corr. with 12 months (*HNZ* 3)

numerical bases for correlation reveals that multiple systems were simultaneously employed in the correlation-building process.

But this concurrent construction by multiple systems was integrated over time, and in this integration, *Wuxing* functioned as a major synthesizing system. One crucial step of synthesis had occurred by the late third century B.C.: the four directions (*Sifang*), the basis for the correlation of other fours, were transformed into five spatial units, with the center added as the fifth direction or *fang*. As seen in the four almanacs (four *Rishu* in Table 3.3), the center and the four directions were treated as five equal spatial units, correlated with dynamic movement in *Wuxing*:[82]

> Metal conquers Wood, Fire does Metal, Water does Fire, Earth does Water, Wood does Earth.
>
> Eastern *fang* is Wood, Southern *fang* is Fire, Western *fang* is Metal, the Center is Earth.

This shift of spatial units from four to five was crucial in the transformation of cosmology from *Sifang* to *Wuxing*, even though most correlations built on fours in these almanacs remained as fours, such as seasons, months, and branches. In this integration of the *Sifang* spatial structure with *Wuxing* cycles, the Center/Earth category was finally established as homogeneous with the other four directions and phases. Now Center/Earth was conquered by East/Wood in exactly the same manner that East/Wood was conquered by West/Metal. This homogeneous nature was, first of all, a complete departure from the Shang and Western Zhou cosmology, in which the center, the absolutely sacred, could never be made equal to the four *fang*. Furthermore, it made possible the total integration of all the correlations of fours and fives into a unified new cosmology.

While the integration of multiple correlative systems appeared in various mantic and military texts in the late third century B.C., it was scholars who completed the large-scale synthesis and construction of a single coherent correlative cosmology. For example, total integration based on five spatial units was seen in two philosophical texts listed in the last two columns of Table 3.3 – *Guanzi* and "Yueling 月令," both products of scholars. *Guanzi* 8 is structured in five spatial units. Without the presence of *Wuxing* terms, most correlations that had been fours are now shifted to the five directions, with the exception of the seasons. A further step of integration, as seen in two other calendars incorporated in *Guanzi* 40 and 41, was to integrate *Wuxing* with these five spatial units.

82 Shuihudi Qimu zhujian zhengli xiaozu, 1990, "*Rishu* A," strips 83b3–92b3, transcription and commentaries on p. 223.

The complete integration of all fours and fives into a uniform system based on *Wuxing* is represented by "Yueling" (Monthly ordinances),[83] the reference in the last column of Table 3.3.

The philosophical synthesis of diverse systems into one coherent correlative cosmology was an enterprise of the scholars to deliberately standardize and theorize cosmology, a means of ordering knowledge in the creation of a new political order and of formulating a cosmological theory of sovereignty for the birth of the unified empire. At the dawn of the imperial era, correlative systems provided a discourse for political debate and resources for imperial ideology, and systematizing cosmology became part of constructing the empire. The Han enterprise of standardizing, moralizing, and canonizing the cosmologies of the Warring States period will be studied in depth in the following chapter. The focus here is the formulation of the cosmological theory of human sovereignty at the end of the Warring States period and during the early Han.

By the third century B.C., as Graham has persuasively demonstrated, philosophers who had once had only a "bare cosmological scheme" – with the building of correlative cosmologies done by other professions outside of philosophy – came to accept the correlative cosmology; and after the unification in 221 B.C., the surviving philosophic schools "took over the whole system of correspondences now indispensable to influence at court."[84] In philosophical texts of this period, such as *Guanzi*, *Lüshi Chunqiu*, and *Huainanzi*, the philosophers adopted the correlative systems of various professions, synthesizing ritual, medical, calendrical, astrological, military, and political discourses to construct a single, all-encompassing cosmology in order to conceive the centralized empire. This synthesized and theorized cosmology, in turn, permeated succeeding professional discourses, consummated in ritual, medical, or calendrical texts. For example, the same calendar for royal rituals, "Yueling," appeared in both philosophical texts such *Lüshi chunqiu* and *Huainanzi* and in the ritual canon *Liji* 禮紀 (Records of rites). Similarly, as Nathan Sivin points out, the medical text *Huandi neijing* brought the cosmological doctrine of *Yin-Yang* and *Wuxing* synthesized in preceding philosophical texts "to its culmination," a doctrine that was "tightly linked to rationales for the Han model of centralized monarchy."[85]

There were three major functions of philosophical synthesis. The first was to build a standardized and coherent correlative cosmology from numerous diverse, sometimes contradictory, correlative systems. A good

83 This text exists in three slightly altered versions found in texts of Qin and Han, primarily in the first twelve chapters of *Lüshi chunqiu*, *Liji*, and in Chapter 5 of *Huainanzi*.
84 Graham, 1986, pp. 91–2. 85 Sivin, 1993, p. 198.

example of such standardization is the royal ritual calendar "Yueling." With its three slightly varied versions in *Lüshi chunqiu*, *Huainanzi*, and *Liji*, dated between the late third century B.C. and the second century B.C., "Yueling" belongs to the genre of almanacs that were popular by the late Warring States period and the beginning of the imperial era, as already seen in the previously discussed *Chu boshu*, *Rishu*, and *Guanzi*, Chapters 8, 40, and 41. Like those almanacs, it prescribes ritual and social actions according to the cosmic order. "Yueling" differs from those almanacs in its systematization, incorporating various sources and standardizing them into a uniform system. While often labeled a "Confucian" text, its cosmological theory was derived from and circulated among a wide range of texts, including texts in other philosophical traditions, mantic practices, and state legal codes.[86]

Such systematization involved incorporating the diagrams, charts, calendars, instruments, and recipes used by various professions into philosophical texts, giving a consistent theoretical interpretation of divergent popular mantic or ritual practices. For example, the correlation of twelve months, ten stems, twelve branches, twenty-eight lunar lodges, and *Wuxing* was the common scheme of the almanacs and the divination instrument called *shi*. Yet the almanacs were assemblages of many different correlative systems, diagrams, and charts, and there was no unifying system to reconcile the discrepancies among different systems. The *shi* instruments used for calendrical divination show a similar discrepancy. The two earliest *shi* instruments unearthed in Hunan province are both dated between 278 and 209 B.C.[87] One of them, from Wangjia tai 王家台, is a square wooden board on which are carved twelve months, twenty-eight lunar lodges, and *Wuxing*. The other one, from Shashi 沙市, is a diagram of two circles, in which are drawn day-signs rather than months, four directions, only four out of five *xing*, and twenty-eight lunar lodges. A third example of *shi*, dated about 167 B.C., includes both the months and day-signs, yet is missing *Wuxing*. It best demonstrates how such instruments were used. It is composed of a round plate on top of a larger square plate[88] (Fig. 3.3).

86 Yang Kuan argues that the various regulations prescribed in "Yueling" derived from the actual policies of various states of the Warring States period, and demonstrates the similarities between the Qin legal codes and "Yueling" regulations. See Yang Kuan, 1980, pp. 62–4.

87 One of the *shi* of the Qin period was from Wangjia tai; see Jingzhou diqu bowuguan, 1985, pp. 42–3. The other is from Shashi, Zhoujiatai. The excavation report of the second *shi* is unpublished. For a description, photograph, and study of the *shi* from Shashi, see Liu Guosheng, 1995.

88 Anhuisheng wenwu gongzuodui et al., 1978. For studies and bibliographies of the *shi* instrument, see Li Ling, 1991a, pp. 5–53; and 1991b, pp. 49–76; Michael Loewe, 1979, pp. 60–85.

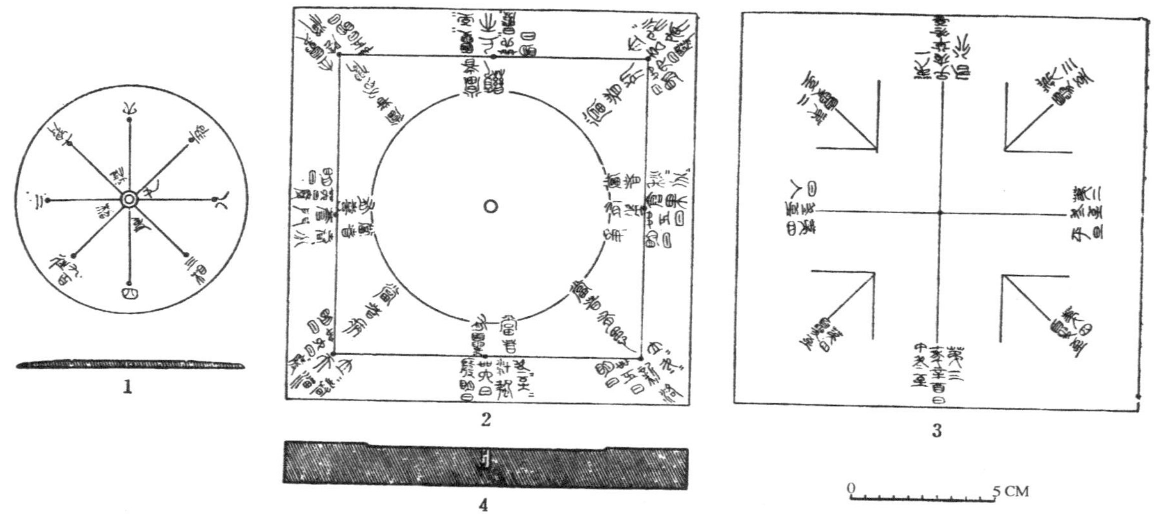

Figure 3.3. *Shi* instrument from the early Han (Anhuisheng wenwu gongzuodui, "Fuyang Shuanggudui Xi Han Ruyinhou mu fajue jianbao," *Wenwu* 8 [1978], p. 25).

The round plate symbolized Heaven, with the northern Dipper in the center surrounded in layers by twelve months and twenty-eight lunar lodges. The square plate symbolized Earth, having three layers of correlation surounding the space covered by the round plate, including ten stems, twelve branches, and twenty-eight lunar lodges. By turning the round plate on top of the square one, seeking a favorable composition, one could determine the auspicious timing for the action concerned.

Other divination diagrams, such as the "Nine-Room Palace Diagram 九宮圖," were also possible sources for the system building of "Yueling." The grand correlative scheme of "Yueling" included a nine-room *Ming tang* 明堂 (bright hall) palace. While there is no archaeological discovery of such a palace before the end of the Former Han, the nine-room palace diagram was popular in divination. "Xingde" B, one of the divination texts discovered in Mawangdui and buried about 168 B.C., has a "Nine-Room Palace Diagram" used for divination[89] (Fig. 3.4).

The diagram is composed of nine rooms (*jiu gong* 九宮), with a round one in the center, four square ones at the four cardinal directions, and four "甲" shaped rooms at points between the cardinal directions. The five *xing* – Water, Fire, Metal, Wood, and Earth – are assigned to the North, South, West, East, and Center.[90] The color symbolism of the Five Phases is also paired with the five *xing* – Earth-yellow, Fire-red, Water-black, Wood-green, and Metal-white. Each of these nine rooms is subdivided into units of twelve, ten, fourteen, and so on, correlated with ten stems, twelve branches, twenty-four solar terms, names of *di* (gods) and celestial deities, and so on. Considering their popularity in daily life, charts like this were likely the sources of the nine-room "bright hall" palace found in "Yueling," although this specific chart from Mawangdui, so elaborate and systematic in its correlation and of a later date then *Lüshi Chunqiu*, may have been influenced by the systematization of the philosophical texts.

When the "Yueling" chapter of the philosophical texts incorporated the correlations appearing in these various almanacs, calendars, *shi* instruments, and divination diagrams, it restructured all the correlations with *Wuxing*. As a result, a total of more than thirty categories were incorporated into this royal calendar, including the correlations used in calendrical, astrological, medical, musical, architectural, bureaucratic, and religious sources. Table 3.4 tabulates the major categories of correlative cosmology synthesized in "Yueling."

89 Fu Juyou and Chen Songchang, 1992, pp. 132–4.
90 Although Earth is not visible in the photograph of the text, that Earth is one of the five *xing* is made clear in the text attached to the chart.

Figure 3.4. The "Nine-Room Palace Diagram" (Fu Juyou and Chen Songchang, *Mawangdui Hanmu wenwu*, Changsha: Hunan chubanshe, 1992, pp. 132–4).

Besides the scale of the system, the philosophical texts also painstakingly resolved the inconsistencies existing among different correlation systems in order to achieve a single, coherent system. They used the *Wuxing* system to reclassify and distribute the correlations originally based on systems of four, six, twelve, and twenty-eight. The four directions, in some *shi* instruments and almanacs, were correlated with four of the five *xing*, with the fifth *xing* sacrificed. In some philosophical texts, such as *Guanzi* 8 and the *Lüshi chunqiu* version of "Yueling," the arrangement was standardized with the center correlated with the fifth *xing*; and the *Huainanzi* 5 version of "Yueling" further transformed *Sifang* and center to five homogeneous spatial positions by giving them the proper name Five Positions (*wu wei* 五位, see Table 3.3). The most difficult incoherence was that between the four seasons and five *xing*; in both *Guanzi* and *Lüshi chunqiu*, this was left unresolved. Yet the *Huainanzi* 5 version of "Yueling" invented five seasons by adding

Table 3.4. *The all-embracing correlative cosmology in early Han*

A. Major "Yueling" correlative categories

Season	Direction	Month	Branches	Stems	28 Lunar lodges*	*Xing*
spring	East	1, 2, 3	*yin, mao, chen*	*Jia Yi*	13th, 15th, 17th	Wood
summer	South	4, 5	*si, wu*	*Bing Ding*	19th, 22nd,	Fire
mid-sum.	center	6	*wei*	*Wu Ji*	24th	Earth
autumn	West	7, 8, 9	*shen, you, xu*	*Geng Xin*	27th, 1st, 4th	Metal
winter	North	10, 11, 12	*hai, zi, chou*	*Ren Gui*	6th, 8th, 10th	Water

No.	Color	Taste	Smell	Notes	Pitch Pipes	Rooms of Palace
8	green	sour	musty	*jue*	*taicu, jiazhong, guxi*	NE, East, SE
7	red	bitter	burnt	*zhi*	*zhonglü, shengbin*	SE, South
5	yellow	sweet	fragrant	*gong*	*baizhong*	Center
9	white	pungent	rank	*shang*	*yizhe, nanlü, wushe*	SW, West, NW
6	black	salty	putrid	*yu*	*yingzhong, huang-zhong, dalü*	NW, North, NE

Thearch	Assistant	Gods	Creatures	Food	Animal	Officials
Tai Hao	Gou Mang	door	scaly	wheat	goat	public works
Yandi	Zhu Rong	stove	feathered	beans	chicken	agriculture
Huangdi	Hou Tu	court	naked	millet	cow	palace revenues
Shao Hao	Ru Shou	gate	hairy	hempseed	dog	military
Zhuan Xu	Xuan Ming	well	shelled	millet	pig	prisons

The work of Heaven and Earth	Ordinance	Catastrophe
birth, sprouting	nourishing, caring	wind, plague of locusts, starvation
growth, blooming	giving, rewarding	drought, fire, plague, dried plantation
flourishing	generosity, kindness	----------
punishing, severity	punishment, chastisement	constant rain, flood, war, plague
death, closing up	funerals, execution	snowstorm, frost, hail, war

B. *Huangdi Neijing*'s expansion of the "Yueling" categories

Human Body	Organs 1**	Organs 2	Emotion	*qi*	Potency	Policy
sinews	liver	eyes	anger	wind	harmony	spreading
blood	heart	tongue	joy	heat	manifesting	enlightening
flesh	spleen	mouth	thinking	moisture	moisten	tranquility
skin/hair	lungs	nose	worry	desiccation	clearance	forcefulness
bone	kidneys	ears	fear	cold	frigidity	cleanliness

* The sequence of the 28 lunar lodges (*xu*) is adopted from *Huainanzi* 3.

** The correlations of organs 1 are also found in "Yueling," as sacrificial organs from animal victims rather than human organs.

"midsummer" to the four seasons, representing the sixth month of the year. Based on the Five Positions and Five Seasons, it correlated all other fours and twelves into *Wuxing*.

The second function of the scholars' philosophical synthesis of diverse correlative systems, besides standardization, was to create a supreme moral authority embodied in a single all-embracing, coherent, theorized cosmology, thus further shifting the authority of interpretation of the cosmic order from various mantic practices to the philosophical texts. The divination texts, instruments, and diagrams were used as practical tools to predict, case by case, the results of an action or the destiny of an individual by locating these actions and subjects in the natural order. Using these tools, a diviner would give an oral interpretation according to the context. Although such divination and other mantic texts had already started to transform an oral tradition into a text-based profession, the philosophical systematization of mantic texts further "decontextualized" the oral discourse, using Jack Goody's term,[91] redefined the meanings of various correlations used in mantic practices, and subjected their meanings to the authority of philosophical theory. Texts such as *Chu boshu*, *Lüshi chunqiu*, and *Huainanzi* commonly attached theoretical essays to the almanacs that they incorporated, loading the correlative schemes with moral meanings. In this process, the meanings of correlations used in mantic practices were extended from simple signs of a determinable natural order to fields of moral debate. Now the cosmic pattern was no longer a simple matter of predicting a good or bad fortune for an individual or the outcome of an event; it had become a matter of the moral foundation for an empire and the model for a new political order. Such theorization and standardization further centralized moral authority from divergent professions in the hands of the possessors of cultural capital – the producers of and commentators upon philosophical texts, and the centralized imperial court that controlled, selected, and canonized those texts.

The third function of this synthesis of correlative schemes, derived from the first two, was the fabrication of a new philosophy of human sovereignty. Based on the grand system and the moral authority of the system, the philosophical synthesis of cosmology addressed a new audience. Instead of serving a diverse population ranging from hegemonic potentates to commoners, as had the mantic practices, the synthesized cosmology addressed the "Son of Heaven," the human

91 Jack Goody defines and describes the "decontextualization" of an oral tradition by writing. He emphasizes that the most important function of transmitting oral discourse by writing is "re-ordering and refining" it in a very different and highly abstract context. See Goody, (1977) 1987, p. 78.

sovereign symbolizing a unified civilization. As shown earlier in this section, physicians, musicians, cooks, and ministers in the court had started to redefine the concept of rulership in their daily interactions with their patrons. The scholars completed this task by creating a comprehensive cosmological philosophy of human sovereignty, defining the ruler in terms of his body, his ritual political behavior, and his social and cosmic function.

In the synthesized correlative cosmology, the ruler's body functioned as the human image of the cosmos, and as an indicator of cosmic movement in the human world. The correlation originally used in almanacs or medical texts for diagnosing sickness now served as the ritual prescription of the ruler's bodily behavior. For example, the correlation of animals (their meat) with directions and colors was found in almanacs as a cause of illness (SHD.*Rishu* in Table 3.2). Yet a similar correlation in "Yueling" became ritual food for the Son of Heaven. Thus the concerns of the correlation shifted from the cause of disease to the ritual behavior of the ruler, from the balance of individual human bodies to the balance of the total cosmic social order through the ruler's body. Not only his food, but also everything about the ruler's body and bodily movements – his clothing, his vehicle, his body ornament, the color of his ritual flags, the musical instruments and the weapons he used, the room in the palace where he resided – were all correlated with and thus completely subjected to the regulation of cosmic cycles of seasons and months. Appropriating the cosmological theory of philosophical texts, *Huangdi neijing* 黃帝內經, a text in the medical genre, combined correlations from ritual and political domains with those used in the medical profession, thus correlating bodily organs, emotions, and temperament with the cosmic movements of *Wuxing*, seasons, and directions, as well as with the policies and offices of government[92] (see Table 3.4).

Such total subjection of the ruler to the cosmic system completed the transformation of the concept of rulership from the hereditary king, whose authority lay in his direct connection to his ancestral spirits, to the new "Son of Heaven," whose authority resided in his role as human representation of the cosmic order. As seen in *Zuozhuan*, minister Zi Chan and physician He had denied any direct contact between the ruler and the spirits by subjecting the ruler's body to the regulation of cosmic patterns. The philosophical synthesis of cosmology further eliminated

92 *Huangdi neijing suwen*, chs. 5 and 67. Nathan Sivin, based on recent studies, suggests that *Huangdi neijing* was compiled of several collections, each containing heterogeneous writings. The writings were done over a century or more and brought together in the first century B.C. See Sivin, 1987, 5n; and 1993.

the ruler's personal, unmediated contact with the world of the divine by subordinating the boundless and indefinite realm of deities and divine powers to the constraints of an all-embracing, coherent correlative system. In earlier sources, such as *Zuozhuan*, *Sunzi bingfa*, and *Chu boshu*, the various gods (*di*), dieties, and spirits were correlated with only a few categories, such as *Wuxing*, directions, and colors, thereby retaining their mystical or magical power because of the small scale of correlation. Once incorporated into the all-embracing correlative system of "Yueling," they became manifestations of the cosmic order rather than unconstrained and incalculable powers. Their will became inessential to the work of the cosmos, and their magical power survived only outside the correlative cosmology in religious practices, such as shamanism and magic, that resisted the cosmological systematization. To contact these powers, the ruler had either to follow cosmic patterns closely, or to use shamans or magicians (*fangshi* 方士) to break the constraints of the cosmological system.

While the synthesized cosmology cut off the ruler's body from direct contact with the divine, it prescribed the day-to-day ritual and political actions of the ruler, subjecting him, as well as the entire human realm he symbolized, to the cosmic order. "Yueling" prescribed that in the spring, the Son of Heaven promulgate the spring ordinance – extending his virtue, bestowing favor, carrying out celebrations and praise, reducing taxes, nourishing the young and the small, protecting the orphaned and childless, opening the granaries to assist the poor – in order to imitate and implement the cosmic phase of growth, nurturing, and production in the human realm. Similarly, in the autumn, the Son of Heaven promulgated the autumn ordinance – searching out and punishing various kinds of unrighteousness, judging criminal cases, increasing the strictness of all punishments, carrying out capital punishment – in order to "encourage the waxing of baleful *qi*" and to imitate Heaven and Earth, which "now begin to be severe."[93] If the Son of Heaven violated such seasonal and monthly ordinances, corresponding catastrophes would surely occur (see Table 3.4).

Conclusion

The transformation of cosmology from *Sifang* to *Wuxing* during the Warring States period was an intrinsic component of the social-political transition from the Bronze Age to the imperial era. It was destructive of

93 *Huainanzi* 5.

the old power relations and of the way power and cosmology had connected, but at the same time was a constructive force, creating new forms of political power, conceptions of social order, and a new conjunction between knowledge and power.

The rising political groups during the Warring States period used correlative cosmologies to deprive the hereditary king of monopolized divine authority and to disseminate it among themselves, beginning to separate divine knowledge from political power. They thus transformed cosmology from a ritual and political structure monopolized by the king to a political discourse accessible to various political forces. The cosmological discourse thus created became an essential means used by various political forces for political debate, for criticism and control of the ruler, and for power contestation during the Warring States period, as well as during succeeding imperial periods. Through such political discourse, power and knowledge separated, battled, and reintegrated, shaping one another in the process of building the new political order of the Qin and Han Empires. These contests reorganized power relations, established sources of moral authority, defined central political issues, and constructed imperial institutions.

As a force destructive of the old political structure, *Wuxing* denied the concept of centrality in *Sifang* cosmology, replacing the sacred and eternal center and the static hierarchy of *Sifang* with cycles of five phases in which cosmic energies constantly interact and transform. Reducing the center to one of the five phases changed it from "the zone of the sacred" to one of the five – whether they are called phases (*xing*), powers (*de*), directions (*fang*), energies (*qi*), or positions (*wei*) – and made it equal to the other four. It was this homogeneous continuum of Center/Earth and the other four *fang/xing* that completely undermined the cosmological and political centrality of the hereditary kings of the Bronze Age. Yet at the same time it initiated an ongoing debate on the position of Center/Earth in cosmology as well as in the new political landscape of a unified empire. The problem of redefining the center in terms of its superiority to or equality with the other four became a focus of contention during the imperial era, reflecting a major political conflict – the conflict between centralization and local autonomy, between unity and separatism, and among various sectors of redistributed power – in the Qin and Han Empires as well as in subsequent empires. As shown in the following two chapters, it took a tremendous amount of effort for the architects of the early empires to relocate the center and restore its supremacy, which was never again the same as it had been during the Bronze Age.

As a constructive force, correlative cosmology established a new form of connection between Heaven and Man – the systematic and comprehensive correlation between the realm of the cosmos or nature and the realm of Man. The building of such an all-encompassing, coherent correlative system accompanied the growth of texts, marking a change in the nature of cosmology from a structure for ritual and political action to a discourse based on texts and systematization.

The systematization of multiple correlative schemes used in mantic practices into a single and coherent cosmology created a supreme moral authority, further shifting the authority of interpreting signs in correlative schemes from the oral discourse of diviners or astrologers to philosophical texts. This systematization and theorization reflects a tension between an immanent and unpurposeful natural order as depicted in mantic practices on the one hand, and the moral intention of Heaven established in philosophical texts on the other. In mantic practices, such as the almanacs or *shi* divination, the gods, spirits (including ancestors of royal clans), and demons were correlated with cosmic patterns of *Wuxing*, seasons, or months, manifesting an immanent, physical cosmic order. As the divination texts and instruments commonly reflect, the success or failure of an endeavor and good or ill fortune for an individual could be diagrammed in a spatial-temporal structure, all predetermined by an inevitable immanent order of the cosmos. One had only to worry about the auspicious timing and direction of one's action, its accordance with the cycles of recession and accretion of the cosmic forces. There seems to have been no room for the moral discussion of one's intention. In this sense, *Wuxing* and other correlative schemes used in mantic practices represent an immanent, unintentional, and almost physical cosmology. But once incorporated into philosophical texts, *Wuxing* became a part of the single, coherent, and purposeful cosmos – Heaven – representing the moral, intentional, and providential aspect of the cosmos as well. The tension between the immanent, unintentional, and mechanical cosmology and a moralized and intentional cosmology of Heaven was to fully unfold during the imperial era of Qin and Han, a subject that will be developed in Chapter 4.

The supreme moral authority of the cosmos created through philosophical synthesis was an intrinsic part of political change. It denied the old rulership by cutting off the ruler's personal contact with the spirits and subjecting his body to the regulation of the cosmos; and it created a new concept of human sovereignty – defining, sanctioning, and restraining the human sovereign of the rising unified empire. The new human sovereign claimed his legitimacy not as the only person who could communicate with his ancestors but as the single exemplary human figure representing the cosmic order to the human world. Yet by the same token, he was utterly constrained by the cosmic order, with his ritual, political, and bodily actions prescribed in the elaborate and systematic royal calendar.

Replacing spiritual mediators, the royal ancestors, with direct correlation between Heaven and Man served to heighten the tension between the political power of the ruler and the moral authority of the cosmos, between the ruler as a responsible agent for cosmic-social order and the prescriber of this order. The king during the Bronze Age was the highest ruler and shaman in one, monopolizing both communication with the divine and the interpretation of divine messages in his one person, with his court shamans as his appendages. During the imperial era, the responsible agent for communication betwen Heaven and Man and the prescribers of cosmic-social order became separated.[94] The resulting tensions during the Han Dynasty between the emperor's interest in establishing personal contact with the infinite divine powers and the constraints of the systematized cosmology, between the political power of the emperor and the moral authority of the scholars, will be discussed extensively in Chapter 5.

94 Benjamin Schwartz has noted that "cosmological Confucianism" had two sides in its relation to the Han Empire. An official orthodoxy reinforcing the imperial polity, it was at the same time used to inhibit and constrain the emperor's despotic aspirations. See Schwartz, 1985, pp. 378–9.

4
Moralizing Cosmology and Transforming Imperial Sovereignty

Introduction

After centuries of change during the Warring States period, China's first imperial era emerged with the Qin unification of 221 B.C. Qin, originally a small state among hundreds of others, ultimately ascended to annex the last six states remaining in a final competition and became the first empire of China. Extremely creative and experimental, Qin lasted only fifteen years. Soon after the death of the First Emperor, the empire was cut short by civil war in 206 B.C., to be replaced by the Han, the first enduring empire of China, which lasted from 206 B.C. to 220 A.D., with only a brief interruption during Wang Mang's Xin 新 Dynasty (9–23 A.D.).

Inheriting Qin's innovations yet making tremendous changes to avoid its short duration, the Han Dynasty accomplished both the institutional and ideological construction of a unified empire, the ideal form of government that served as a model throughout China's later history. Such a model comprises the institution of emperorship, the concept of imperial sovereignty, the centralized bureaucratic government, a theory of dynastic succession, and the moral authority legitimizing imperial rule. Although often troubled by internal deficiencies, and periodically challenged by foreign invaders, the imperial model established by the Han Dynasty was held authoritative, by both Chinese and foreign rulers of succeeding dynasties, until the European intrusion of the nineteenth century.

In the institutional and conceptual construction of the empire during the Han, *Wuxing* cosmology became a shared political discourse. Besides being held as a cosmological model for imperial sovereignty and the bureaucracy, *Wuxing* also became the predominant cosmology used for political dispute. It was the primary cosmological system employed in omen interpretation, prescribing a particular cosmic-social order and condemning any deviation from it. Thus *Wuxing* was not a homogeneous orthodoxy that simply mystified, legitimated, or justified the existing political order. Rather, it was a dialectic cosmology with which power was constituted and contested. Within *Wuxing* cosmology, competing social forces combatted over the sociopolitical order, the transmission of power, and the issue of imperial sovereignty.

To illustrate *Wuxing* cosmology as such a dialectic political discourse in the Han Dynasty and its essential position in the construction of imperial institutions and ideology, this chapter analyzes the most comprehensive representation of this discourse – "Wuxing zhi 五行志" (The treatise on the Five Phases) in *Hanshu* 漢書 (The history of the Former Han Dynasty). In treating some seemingly textual or structural problems of the treatise, this chapter reveals the Han moralization of cosmology as it was inherited from the Warring States and Qin, and the transformation of the concept of imperial sovereignty in the first two centuries of the imperial era, from that exemplified by the First Emperor of Qin to that used by Wang Mang 王莽 (45 B.C.–23 A.D.), the epitome of Han Confucian ideology.

"Wuxing Zhi" in *Hanshu*

The Content of "Wuxing Zhi"

"Wuxing Zhi" (The treatise on the Five Phases) in *Hanshu* belongs to one of the four major genres of history writing – *zhi* 志 (treatises or monographs) – that were invented during the Han Dynasty. Such treatises dealt with specific subjects that were significant to the government. As Etienne Balazs has pointed out, treatises, and histories as a whole, were "written by officials for officials" and conceived as guides to administrative practice. The particular value of treatises was that they were aimed at producing "not scholars, but statesmen and administrators who were knowledgeable about all government activities."[1]

Using *Wuxing* in its title, "Wuxing zhi" in *Hanshu* was devoted to "the Way of Heaven and Man," the cosmic-social order verified through the resonance between Heaven and Man, and manifested in various abnormal phenomena in the universe. Far from being a coherent theory of a single author, it represented a complex cosmology formulated and disputed by the major theorists of the twelve regimes of the Former Han, including Dong Zhongshu 董仲舒 (ca. 179–104 B.C.), Liu Xiang 劉向 (79–8 B.C.), Liu Xin 劉歆 (d. 23 A.D.), Jing Fang the Younger 京房

1 Balazs, 1964, pp. 129–49.

(d. 37 B.C.),[2] and many others. It records 378 cases of calamities and prodigies, with various interpretations of them by the major theorists as portents or omens of disorder in government.

Ban Gu 班固 (32–92 A.D.), the author of *Hanshu* and the forefather of China's dynastic histories, compiled this text, thereby creating the genre of *Wuxing* for treatises.[3] Ban Gu also gave *Wuxing* cosmology prime importance in Han historiography, making it the longest of all ten treatises and the most voluminous of all subjects in *Hanshu*. Following Ban Gu's invention, *Wuxing* as a genre of treatise was adopted by the standard histories. Seventeen of the twenty-five standard histories included the genre of treatises, while the eight others did not have treatises at all.[4] Among the seventeen dynastic histories with treatises, thirteen had a treatise on *Wuxing* or omens, continuing the practice of *Hanshu*. Even when the Han Confucian tradition was under the most serious attack by

2 There were two men named Jing Fang in the early Han Dynasty. A. F. P. Hulsewé has clarified the distinction between them, concluding that Jing Fang the Elder lived roughly between 140 and 80 B.C., and that Jing Fang the Younger, whose original surname was Li, was beheaded in 37 B.C. at the age of forty. Hulsewé believes that it was Jing Fang the Elder who was the author of the the work known as the *Jing Fang Yizhuan*, quoted many times in "Wuxing zhi" of *Hanshu*. See Hulsewé, 1986.

But I am inclined to believe that it was Jing Fang the Younger who was quoted in "Wuxing zhi" of *Hanshu*. This belief is supported by three kinds of evidence. First, the biography of Jing Fang, in chapters 75 and 88 of *Hanshu*, describes a Jing Fang the Younger, whose original name was [Li] Mingjun, who studied with Jiao Yanshou, a claimed student of Meng Xi. It was also this Jing Fang who was promoted to high positions in the central government because of his ability at interpreting omens, and who was eventually beheaded under persecution by Shi Xian. And according to Liu Xiang and Ban Gu, Jing Fang the Younger developed a learning of *Yi* that was different from all others, so there was a tradition of Jing's learning of *Yi*. (*HSBZ* 88, pp. 10a–b; *HS*, p. 3601.)

Second, the system of *Jing Fang Yizhuan* as quoted in "Wuxing zhi" accords with the biographical description of Jing Fang the Younger's system. A systematic reconstruction of these quotations in "Wuxing zhi" and a comparison of such a reconstruction with other works under the title *Jing Fang Yizhuan*, including *Jing shi Yizhuan* in *Han Wei congshu*, needs to be done.

Finally, in "Wuxing zhi" of *Hanshu*, Ban Gu treats Jing Fang the Younger as the author of the *Jing Fang Yizhuan* he quotes. In the case of a hen transformed into a cock, dated 49 B.C., Ban Gu quotes *Jing Fang Yizhuan* as interpreting this case as a sign predicting the death of the ones who can predict the timing of things. Ban Gu follows by saying that Jing Fang believed it was he himself who would die because he knew how to predict timing. (*HSBZ* 27Ba, pp. 12a–13b; *HS*, pp. 1370–1.)

3 The conclusion that "Wuxing zhi" was compiled by Ban Gu is supported by two pieces of evidence. First, the antecedent of *Hanshu*, *Shiji*, from which Ban Gu adopted a large amount of material, does not have a treatise on *Wuxing*. Secondly, after Ban Gu's death, only the treatise on Patterns of Heaven and the Eight Tables (*biao*) were known to be incomplete and were finished by Ban Chao and Ma Rong. For this evidence, see the biography of Ban Chao in *Hou Hanshu*. This conclusion is also supported by A. F. P. Hulsewé; see Hulsewé, 1993.

4 These eight histories without treatises are histories of the dynasties during the times of disunity.

the Song and Ming intellectuals, the treatise on *Wuxing* as a genre concerning the resonance between Heaven and Man was accepted and maintained. For example, the authors of *Xin Tangshu* 新唐書 and *Yuanshi* 元史 both criticized the Han Confucian tradition of interpreting omens. *Xin Tangshu* reprimanded Han Confucians such as Dong Zhongshu, Liu Xiang, and Liu Xin, saying that they all "missed the original meanings of the sages" as stated in *Spring and Autumn* and in *Hongfan* 洪範 (The great plan).[5] *Yuanshi* also maintained that Han Confucians "did not understand the great principle" of resonance between Heaven and Man established by the ancient sages.[6] Yet authors of both the histories continued to compile a treatise on *Wuxing*, acknowledging that "Man, Heaven and Earth resonate with one another as three poses, and omens thus rise responding to the proper categories,"[7] and that "the grave catastrophes and omens between Heaven and Earth are caused by disorder in government."[8] Table 4.1 demonstrates this continuing influence of Ban Gu's creation of "Wuxing zhi" as a history genre up to the twentieth century.

The statistics presented in Table 4.1 illustrate that except for *Weishu* and *Qingshigao*, which replaced *Wuxing* with "omen" (as *lingzheng* 靈徵 in *Weishu*, and as *zaiyi* 災異 in *Qingshigao*), all other histories continued the title "Wuxing zhi." The position of "Wuxing zhi" in the category of treatises indicates not so much the changing importance of "Wuxing zhi," but the changing position of the larger category to which "Wuxing zhi" belonged. Balazs has categorized all treatises into four groups – rites, "sciences," government institutions, and bibliography. Balazs also observes a trend toward decreased attention to rites and "sciences" and increased attention to institutional matters, based on the amount of space occupied by each category.[9] Besides this shifting attention, however, it is worth noticing that after *Jinshi* 金史, "sciences" – which group "Wuxing zhi" together with astronomy and the calendar – moved from a secondary to the first category, occupying the primary position of all treatises, while the rites category became secondary. It was for this reason that "Wuxing zhi" rose to be the second or third of all treatises, located between astronomy and geography – in other words, between the treatise on Heaven and that on Earth.

With its long-enduring influence in standard histories, "Wuxing zhi" in *Hanshu* has also been criticized throughout history and into modern times. There have been two major criticisms of "Wuxing zhi." The first is of its "falsity." Because Ban Gu made cosmology a crucial component

5 *Xin Tangshu*, p. 872. 6 *Yuanshi*, p. 1049. 7 Ibid. 8 *Xin Tangshu*, p. 872.
9 Balazs, 1964, p. 140.

Table 4.1. *The "Treatise on the Five Phases" in the twenty-five standard histories*

Title	Title of the treatise	Position	Structure of the treatise
Hanshu	Wuxing zhi	no. 7 of 10 zhi	Five *xing* of Earth, Five Duties of Man, Heaven
Hou Hanshu	Wuxing zhi	no. 5 of 8 zhi	Five *xing* each combined with one of the Five Duties, Heaven
Songshu	Wuxing zhi	no. 6 of 8 zhi	Five *xing* each combined with one of the Five Duties of Man
Nan Qishu	Wuxing zhi	no. 7 of 7 zhi	Five *xing* each combined with one of the Five Duties of Man
Weishu	Lingzheng zhi	no. 7 of 9 zhi	Classifying omens by kind
Jinshu	Wuxing zhi	no. 9 of 10 zhi	Five *xing*, Five Duties of Man
Suishu	Wuxing zhi	no. 5 of 10 zhi	Five *xing*, Five Duties of Man
Xin Tangshu	Wuxing zhi	no. 6 of 13 zhi	Five *xing*
Jinshi	Wuxing zhi	no. 3 of 14 zhi	Classifying omens chronologically
Songshi	Wuxing zhi	no. 2 of 14 zhi	Five *xing*
Yuanshi	Wuxing zhi	no. 2 of 13 zhi	Five *xing*
Mingshi	Wuxing zhi	no. 2 of 16 zhi	Five *xing*
Qingshigao	Zaiyi zhi	no. 2 of 16 zhi	Five *xing*

of historiography, he was later criticized for his "falsity." Liu Zhiji 劉知己 (661–721) dismissed this treatise altogether as false analogy, rejecting the notion that the Way of Heaven is connected to human affairs.[10] Primarily based on this treatise, modern scholars such as Ran Zhaode and An Zuozhang conclude that Ban Gu wrote *Hanshu* to propagate "mysticism" or an "apocryphal theology."[11] These critics judge "Wuxing zhi" anachronistically according to their own rationality, rather than placing it in its historical context. They also confuse a work of historiography with a writer's personal intellectual statement. "Wuxing zhi" is a piece of history rather than a theoretical essay.[12] It is the significance of *Wuxing* cosmology in Han political life and the treatise's rich representation of it that count for the value of "Wuxing zhi," rather than its rationality according to later standards.

The second criticism of "Wuxing zhi" has been its "incoherence." "Wuxing zhi" represents the disputing, contesting, and transforming of *Wuxing* cosmology rather than a coherent theory. The diversity, contradictions, and inconsistency in the *Wuxing* cosmology it represents, however, became the other focus of criticism of its compiler. Liu Zhiji devoted a whole volume of his *Shitong* (Critiques on histories) to the critique of "Wuxing zhi." He pointed out four major flaws in this treatise, all relating to the incoherence, irrationality, and self-contradictions of the text.[13]

Though responsible for creating this treatise in the imperial history, Ban Gu nevertheless was not the "author" of "Wuxing zhi" but rather its compiler. "Wuxing zhi" is a synthesis of heterogeneous texts on the subject. As Balazs has pointed out, "treatises are a mosaic of texts and extracts of texts, of actual passages from calendars, astronomical and mathematical calculations . . . together with passages from innumerable memorials, requests, and petitions."[14] This is particularly true of "Wuxing zhi" in *Hanshu*. The purpose of compiling this treatise, as Ban Gu clearly stated in its introduction, was precisely to distinguish the different voices represented in divergent sources, and thus to represent the complexity of this predominant political discourse rather than to present one coherent view:[15]

10 Liu Zhiji, *Shitong*, 1978, vol. 3, p. 62.
11 Ran Zhaode, 1962, pp. 33–8; An Zhuozhang, 1979, p. 93.
12 Past scholarship has established the reliability of Ban Gu's *Hanshu* as a work of high historical accuracy. See Homer H. Dubs, 1946, pp. 23–43; Hulsewé, 1961, pp. 31–43.
13 Liu Zhiji, *Shitong* 19; see *Shitong tongshi*, rpt. 1978, pp. 533–70.
14 Balazs, 1964, p. 141.
15 *Hanshu buzhu* 27A (abbreviation *HSBZ*), Wang Xianqian, 1983, pp. 2a–b; *Hanshu* (abbreviation *HS*), Ban Gu, (1962) 1987, p. 1317. Translations of "Wuxing zhi," *HS* 27, are my own.

> The Han was established, inheriting the condition after the extinction of scholarship by the Qin. During the regimes of Emperors Jing and Wu, Dong Zhongshu studied the *Gongyang Spring and Autumn* 公羊春秋, and initiated the reasoning from *Yin-Yang* [analogy], becoming the forefather of the [Han] scholars. After the regimes of Emperors Xuan and Yuan, Liu Xiang studied the *Guliang Spring and Autumn* 穀梁春秋, accounted [its recorded] good and bad fortunes and commented on them according to "Hongfan," [with his theories and interpretations] conflicting with Dong Zhongshu. Afterward, Liu Xin, Xiang's son, studied the *Zuo Commentary* [on *Spring and Autumn*], with once again a deviant understanding of *Spring and Autumn*, and a different explanation of the *Commentary on Wuxing* 五行傳. Accordingly, I represent Dong Zhongshu, distinguish Liu Xiang and Xin, and record in the biographies the accounts of events by people of the same type such as Sui Meng 眭孟, Xiahou Sheng 夏侯勝, Jing Fang, Gu Yong 谷永, Li Xun 李尋, enumerating [the manifestation of this doctrine in] the twelve reigns up to that of Wang Mang, in order to add on to *Spring and Autumn*. I am hereby writing this text.

During the time of the "twelve reigns" of the Former Han Dynasty, most political debates were carried out within *Wuxing* cosmology. It is precisely the preservation of such debates and contradictions in "Wuxing zhi," at the price of "self-contradiction" and "incoherence," that is extremely valuable for understanding the political contest carried out in the *Wuxing* discourse. The following sections of this chapter, therefore, will analyze two of the contradictions observed in the structure of the text, using them to disclose the political tensions and conflicts involved in the building of the Han Empire.

The Structure of "Wuxing Zhi"

"Wuxing zhi" derived its structure from "Hongfan" (The great plan), one of the cardinal chapters of the corpus of canons used in Han orthodoxy.[16] "Wuxing zhi" classifies its 378 cases of omens in categories borrowed from "Hongfan," introducing each subsection with a quotation from "Hongfan" and its earlier commentaries. As much as it appears to be an elaboration of the canon, however, "Wuxing zhi" made major changes in the canon's structure. These structural changes are not merely textual issues, but rather landmarks in the transformation of cosmology and the concept of imperial sovereignty.

"Wuxing zhi" opens with a quotation from "Xici 繫辭," an appendix to the *Book of Changes* (*Yijing* 易經):[17]

16 "Hongfan," a chapter of *Shangshu*. For a recent study of this chapter, see Michael Nylan, 1992.

17 *HSBZ* 27A, p. 1a; *HS*, p. 1315; references for "Xici" of *Yijing* are to Kong Yingda, *Zhouyi zhengyi*, in *Shisanjing Zhushu*, vol. 1, ed. Ruan Yuan, rpt. 1980.

Heaven sent down images showing good and ill fortunes,
The sages read the images.
The [Yellow] River delivered the Diagram.
The Luo River delivered the Document.
The sages took them as models.

Opening with this quotation, Ban Gu means to trace the divine origin of the sacred canons upon which his treatise as well as the whole *Wuxing* discourse are based. To do so, he then quotes Liu Xin, who *says* that the first human king, Fu Xi 伏羲, received the River Diagram (*Hetu*) from Heaven, and by imitating the diagram derived the Eight Trigrams of the *Book of Changes*. Then the mythical King Yu 禹 was given the Luo Document (*Luoshu*), which he elaborated into "Hongfan." Liu Xin further identifies sixty-five characters from "Hongfan" as the original Luo Document, which lists the nine categories.[18]

As much as they held the canon as the authority of the divine and antiquity, Ban Gu and the authors of the texts that he incorporated into "Wuxing zhi" nevertheless made two fundamental changes from the structure of "Hongfan." The first change was in the sequential ordering of *Wuxing*. After the introduction tracing the divine origin of the canons, the main text of "Wuxing zhi" begins by quoting "Hongfan," ordering *Wuxing* in the sequence of Water-Fire-Wood-Metal-Earth. This sequential ordering of *Wuxing*, as Chapter 3 has discussed, was a result of imposing *Wuxing* on an archaic *Sifang*-center spatial order concerning the ruler and his position in a political space (see Chapter 3, Table 3.1), and it was from this sequence that the conquest cycle of *Wuxing* derived.

After this citation from "Hongfan," however, "Wuxing zhi" starts its explanation with the third *xing*, Wood, which initiates a generation cycle of Wood, Fire, Earth, Metal, and Water (generation cycle 1 of *Wuxing*, see Chapter 3, Table 3.1). The sequential order of "Hongfan," together with the concerned political space, is thus replaced by the generation cycle of *Wuxing*. With no explanation given by Ban Gu or later commentators on "Wuxing zhi," this change is made in an abrupt fashion that has puzzled modern scholars.[19]

The second structural change is the incorporation of *Wuxing* into a cosmology of Earth, Man, and Heaven. Although based on "Hongfan" and titled *Wuxing*, "Wuxing zhi" as a whole is structured according to neither the nine categories of "Hongfan" nor the five categories of

18 *HSBZ* 27A, pp. 2b–3a; *HS*, p. 1316.
19 W. Eberhard explains the order of *Wuxing* adopted by Ban Gu, but does not explain why this order differs from the citation from "Hongfan." See Eberhard, 1970; Beck, 1990, p. 133.

Wuxing, but rather according to the three realms, those of the Earth, Man, and Heaven. The 378 cases of omens are divided into these three realms, with those manifested on Earth correlated to *Wuxing*, those manifested in human beings correlated to the Five Duties (category 2 of "Hongfan"), and those manifested in Heaven (astronomical portents) correlated to the "sovereign standard" (*Huang ji* 皇極, category 5 of "Hongfan").

These two changes show a contradiction in the treatment of the canon by the compiler and authors of "Wuxing zhi." On the one hand, they cite "Hongfan" as the divine blueprint for their enterprise of illuminating the "Way of Heaven and Man." On the other hand, they find it necessary to replace the "Hongfan" order of *Wuxing* with the generation cycle that they choose, and to substitute their cosmology of Earth, Man, and Heaven for the nine-category structure of "Hongfan." In changing the basic structural order of a canon that they themselves held as sacred, the compiler and authors did not simply reorganize the structure of a text. Rather, through manipulating cosmological metaphors, they were exercising their power over the definition of imperial sovereignty and of the political, moral, and cosmological order of the human world.

The following sections analyze these two changes in structural principles, demonstrating how a major political contest over imperial sovereignty was manifested in the cosmological metaphors.

Conqueror or Nurturer: Moralizing the Cosmos and Debating Imperial Sovereignty

The first structural change in the canon, the replacement of the Water-Fire-Wood-Metal-Earth sequence of "Hongfan" with a generation cycle of Wood-Fire-Earth-Metal-Water, marked an ongoing battle over imperial sovereignty. As seen in Chapter 3, various *Wuxing* cycles, including conquest, generation, and the magic square cycles seen in "Hongfan," were already in use during the Warring States period. But in the formation of the Qin and Han Empires, *Wuxing* and its various cycles were employed to define imperial sovereignty and to explain the dynastic succession. Therefore, the debate over choosing one cycle or another as the orthodoxy epitomizes the disputes over the legitimization and transmission of power, the foundation of imperial sovereignty, and the way of government.

These political disputes were cast primarily in the form of choosing a dynastic symbol from the *Wuxing* system. Over the first 246 years of imperial history, from the Qin unification (221 B.C.) to the founding of the Later Han Dynasty (25 A.D.), this symbol had been changed twice, from Water to Earth to Fire. Each change was accompanied by a change in a set of institutions and ritual protocols, including calendars, ritual colors, numbers, robes, music notes and instruments, names of offices, and even policies of government (for example, law and punishment versus reward and education). The debate over choosing the right symbol engaged most influential statesmen and scholars, and cost some of them their lives and careers. The three symbols were installed by the three most ambitious emperors in the era of the first empires, namely the First Emperor of Qin; Emperor Wu, whose reign marked the fundamental changes in ideology and state policy during the Former Han; and Emperor Guang Wu, the founder of the Later Han.

The change in the dynastic symbol from Water to Earth and then to Fire signals a moralization of *Wuxing* cosmology, a shift from a conquest cycle to a generation cycle as the basis for imperial sovereignty and the transmission of power. Former scholarship has established that these changes in the symbol of the dynasty were means of legitimization of imperial sovereignty.[20] Building on this scholarship, the following discussion discloses how such legitimization actually embodied intense political contestation, and how different symbols represented conflicting models of imperial sovereignty – that of a world conqueror versus that of a world nurturer.

Water and the Conquest Cycle of the Five Powers

Water as the symbol of the dynasty after Zhou derived from the conquest cycle of *Wuxing* and was used by the First Emperor to justify both the transmission of power by force and violence and an imperial sovereignty based on military accomplishment, discipline, and punishment. Soon after the Qin conquered the other six states and became the unifier of "all under Heaven," the First Emperor embarked on the cosmological legitimization of imperial sovereignty by adopting Water as the dynastic symbol. *Shiji* 史記 records the theoretical foundation of the First Emperor's adoption of Water as the symbol:[21]

After the First Emperor of Qin had just unified all under Heaven and proclaimed himself the emperor, someone advised him, saying: "When Huang Di [the Yellow Emperor] received the Power of Earth, Yellow Dragon and Earthworm appeared. When the Xia received the Power of Wood, Green Dragon

20 The most thorough studies of how *Wuxing* theory was used to legitimize imperial power include Gu Jiegang, (1926–41) 1982, vol. 5, pp. 404–616; Loewe, 1979a.

21 *Shiji huizhu kaozheng* 28, pp. 19–20 (abbreviation *SJHK*); *Shiji*, p. 1366 (abbreviation *SJ*).

Table 4.2. *The Five Powers theory of Zou Yan*

Ruler	Power	Omen	Color
Huang Di (Yellow Emperor)	Earth	Grand earthworm and mole cricket	Yellow
Yu (of Xia)	Wood	Luxuriant vegetation in the winter	Green
Tang (of Shang)	Metal	Blade appearing in the water	White
King Wen (of Zhou)	Fire	Red birds carrying red book	Red
The succeeding dynasty	Water	The *qi* of water being predominant	Black

Source: *LSCQ*

stopped at the suburb and the grass and trees were luxuriant. When the Shang received the Power of Metal, silver flowed out of the mountains. When the Zhou received the Power of Fire, there was an omen of red birds. Now the Qin has replaced the Zhou, and the time of the Power of Water has come."

In traditional histories, the theory of dynastic transmission based on the rotation of the Five Powers has been attributed to Zou Yan. Even though Zou's works have long been lost, fragmentary accounts of his theory have been the basis for this belief. A paragraph from *Lüshi chunqiu* is commonly considered to be a preserved fragment of Zou's work.[22] It states: "Whenever a sovereign is going to rise, Heaven will certainly show favorable omens to the people in advance." It then describes the dynastic transmission and the symbols of each dynasty. The system is summarized in Table 4.2, which shows a system basically identical to the one suggested to the Qin emperor.

The records from *Shiji* and from *Lüshi chunqiu* are clearly two versions of the same theory. Both attribute the transmission of dynastic power to the circulation of the Five Powers in a conquest cycle – Earth is conquered by Wood, Wood by Metal, Metal by Fire, and Fire by Water – with each power verified by its correlating omens and color; and both describe human history in the first cycle of *Wuxing*, which starts with the Yellow Emperor's Power of Earth and is completed by the Power of Water. *Shiji* clearly states that it was this theory of Zou Yan that the First Emperor of Qin adopted:[23]

22 *Lüshi chunqiu* 13, pp. 4a–5b, SPPY edition (abbreviation *LSCQ*).
23 *SJHK* 28, p. 23; *SJ*, p. 1368.

The disciples of Master Zou compiled *Zhongshi Wude zhi yun* 終始五德之運 (The course of the beginning and end of the Five Powers). Up to the time of the Qin Emperor, the Qi people submitted this book and the First Emperor put it into practice.

Although the First Emperor of Qin largely copied Zou's theory, there is a conflict between Zou's theory and Qin's practice that has hardly been discussed, that is, the conflict between the moral authority of the cosmology and the political power of the emperor.

By incorporating *Wuxing* cosmology used in mantic practice with Confucian ethics, Zou Yan was trying to confine political power within ethical boundaries. *Shiji's* biography states that Zou Yan created his theories because he saw that "the rulers were wallowing in luxury and pleasure and failed to uphold morality." Zou's theories, though "extravagant and unorthodox," were nevertheless "bound to submit to benevolence, righteousness, regulation, and frugality. Thus Zou initiated the regulation of conduct of the ruler and the subjects, from the top to the bottom [of the hierarchy] and within the six human relations."[24] Zou Yan, therefore, was one of the late Warring States scholars who tried to confine political power within the moral authority of cosmology. By claiming that the new ruler of a unified empire must receive the mandate of the cosmos, which Heaven sends down omens to verify, Zou was conveying the message that the rulership was not based on sheer force or the ruler's personal ambition, but was to be installed by the moral authority of Heaven and cosmic destiny. Rhetoric of this kind had been used to restrain the political power of the hegemonic kings before Zou Yan. Duke Huan of Qi made himself "hegemon" (*ba* 霸) and wished to carry out the *feng* 封 and *shan* 禪 rituals in claiming the Mandate of Heaven. His justification was his military and political domination of all under Heaven, which, he believed, differed not at all from the mandate enjoyed by the Three Dynasties – Xia, Shang, and Zhou. Guan Zhong, however, restrained the duke's ambition by arguing that the mandate had to be verified not by political and military power but by auspicious omens from Heaven.[25]

While the First Emperor of Qin put Zou's cosmological system into practice, he nevertheless recreated the meaning of *Wuxing* and therefore reversed Zou's original intention. He installed the following institutions in the name of *Wuxing*:[26]

Calculating the transmission cycle of *Wude* 五德 (the Five Powers), the First Emperor held that Zhou had the Power of Fire, and Qin was replacing the Power

24 *SJHK* 74, pp. 5–7; SJ, p. 2344. 25 *SJHK* 28, pp. 12–15; *SJ*, p. 1361.
26 *SJHK* 6, pp. 23–5; SJ, pp. 237–8.

of Zhou and should follow the Power that Fire could not conquer. From then on the Power of Water had begun its dominance. Let the beginning of the year be changed and all tributes to the court start at the eleventh month; let robes, garments, feathers, fur, stakes, and flags all be in black; let the number six be used for regulations . . . Let the [Yellow] River be re-named the De Water, symbolizing the beginning of the Power of Water. [Let the policy] be harsh, firm, perverse, and occult, with all affairs determined by law. Be severe and strict rather than benevolent (*ren*) 仁, kind (*en*) 恩, harmonic (*he*) 和, and righteous (*yi*) 義. Only then will it be in accordance with the number of the Five Powers.

Here the function of *Wuxing* and auspicious omens changed, from providing sources of moral authority restraining political power to legitimating despotism. Symbolizing death and punishment, Water justified the imperial foundation of the military and punishment. The color and number associated with Water were used in ceremonies, confirming political power. Changing the calendar implied the correlation of the dynasty with the temporal order of Heaven, and changing the name of the Yellow River claimed the imperial sovereignty on Earth.

The conquest cycle of *Wuxing* cosmology legitimated this very kind of sovereignty and transmission of dynastic power. Qin's assuming sovereignty from Zhou only manifested the cosmic principle that Water conquers Fire, and Qin conquering Zhou by means of sheer force was seen as the fulfillment of destiny. The highest Confucian moral principles to which Zou's theory submitted – benevolence, kindness, righteousness, and harmony – were specifically eliminated, because they went against the symbolism of Water.

The First Emperor of Qin thus defined imperial sovereignty as a despotism based on force and violence with which he achieved political unification. His proclaimed title – "Huangdi 皇帝" (the August Emperor) – was one way of glorifying his personal military achievement:[27]

I with my single body raised the troops to put down the revolt and disorder. Thanks to the help of ancestral spirits, I convicted the six kings for their crimes, and put all under Heaven to order. Without changing the title of the throne, there would be no way to glorify these achievements and pass them on to the later generations.

To the First Emperor of Qin, such absolute political power of conquest had to be free from the control or contradictory influence of any other authority. Among many changes to political institutions, the First

27 *SJHK* 6, p. 21; *SJ*, p. 236.

Emperor of Qin particularly abolished the Zhou institution of *shi* 謚, a system of naming deceased rulers according to their achievements in life, because no evaluation or criticism of the ruler by the subjects could be allowed.[28] While using scholars in his court, he never submitted himself to the moral authority they claimed to have, and the conflict between moral authority and political power soon broke out into bloodshed. When the scholars used antiquity and Heaven to criticize the regime, the emperor ordered a wholesale burning of books and burying of scholars.[29] The emperor further made rituals an extension of his power. He not only dismissed scholars from his grand *feng* and *shan* rituals, keeping documents confidential between himself and the god,[30] but also turned his frequent ritual journeys into eulogies of his personal power and achievement, with hardly any credit given either to divine powers or to collective effort in assisting his success.[31]

Therefore, the First Emperor used the Five Phases to justify his type of sovereignty, a sovereignty gained by conquering the world and one that ruled by force. This sovereignty was realized in a centralized totalitarian government and was based on economic strength, military expansion, extravagant rituals, and discipline by reward and punishment. Its consequences were the regime's exhaustion of material and human resources and its short lifespan of fifteen years.

Throughout Chinese history and in modern scholarship as well, the Qin emperor has been regarded as an extreme case of despotism, an absolute power and the great unifier of China, just as he claimed to be. But cosmology also placed constraints on him that have long been overlooked. Zou Yan's theory of the succession of Five Powers was not invented to flatter the ruler, but rather was adapted from a *Wuxing* discourse widely used in social practice in order to imply a divine authority that was above political power. While the First Emperor could turn this common discourse into a justification of his sovereignty, he could

28 *SJHK* 6, pp. 22–3; *SJ*, p. 236. 29 *SJHK* 6, pp. 49–58, *SJ*, pp. 254–8.

30 In order to perform the grand *feng* and *shan* sacrifices to claim the mandate, the first emperor of Qin summoned seventy Confucian masters and scholars from Qi and Chu to meet with him at the foot of Mount Tai. But these scholars talked about how in ancient times the *feng* and *shan* sacrifices were kept humble and easy, even "the wheels of the carriages were wrapped in rushes to avoid doing any injury to the earth and grass of the mountain." The First Emperor was surely repelled by these restrictions. He therefore dismissed all the scholars, had a carriage road opened up, ascended to the summit, and set up a stone marker "extolling his own achievement as the First Emperor." He also kept the documents used in the sacrifices strictly secret, and no men of the time were able to record these ceremonies. *SJHK* 28, pp. 20–21; *SJ*, pp. 1366–7.

31 Six eulogies as such were inscribed on stone during the First Emperor's regime and were recorded; see *SJHK* 6, pp. 32–65; *SJ*, pp. 242–262.

not avoid the fluid meaning of the cosmology, which permitted conflicting interpretations. Nor could he resolve the contradiction between this cosmology and his desire for an eternal emperorship: "I shall be called the First Emperor, and succeeding generations of emperors shall be counted by the number – the Second, the Third, and so on up to the ten thousand generations – and the succession passing down forever."[32] While the conquest cycle of the Five Phases supplied his sovereignty with a cosmological justification, it also predicted its demise, since Water was doomed to be conquered by Earth.

The obvious contradiction between the succession of Five Powers and the First Emperor's desire to have his empire passed down forever has puzzled some great scholars such as Derk Bodde. Bodde, trying to explain the contradiction, tends to believe that the system of Five Powers used in the edicts of the First Emperor of Qin was not authentic, but rather was inserted into historiography by later historians.[33] But attributing an ideological contradiction to textual corruption cannot be conclusive, particularly because it sacrifices historical complexity for the sake of logical consistency. It is my argument that precisely this logical contradiction illuminates the fundamental dynamism of Qin and Han history, that is, the conflict between the moral authority of the scholars and the political power of the throne. This dynamism was one of the major driving forces not only in the downfall of the Qin, but also in the formation and transformation of the Han Empire.

Earth and the Conflict between Wuxing and the Three Unities

The Han Dynasty took over the imperial sovereignty of the Qin and ruled (with a break in the middle) for about four centuries. As a result, Han has remained the model of political unity and stability in China. *Wuxing*, synthesized with other cosmological systems into a single coherent cosmology that pervaded every aspect of Han culture by the middle of the dynasty, represented a cultural unity that echoed this political unity. But *Wuxing* became so pervasive only because it served as a shared discourse through which competing forces fought for their own political goals.

The two most important competing political forces have been described by Michael Loewe as "modernists" and "reformists," with each category identified by a shared attitude and set of policies.[34] Loewe

32 *SJHK* 6, p. 23; *SJ*, p. 236.
33 For a summary and evaluation of this judgment, see Bodde, 1986, pp. 77, 96–7.
34 For a detailed description of these two forces and their conflicts, see Loewe, 1986b, pp. 103–6.

describes the shift from "modernist" to "reformist" policies as dominant at court, a shift that occurred around 81 B.C. While these two categories do not precisely coincide with specific groups, I have observed that the "modernists" were often high statesmen of the central government whose authority came from their political positions and their achievements in building the economic, military, and political strength of the empire. Representatives of modernist statesmen include Huo Guang 霍光 (d. 68 B.C.), a relative of the empress of Emperor Wu and marshal of state (*dasima* 大司馬, also translated as commander-in-chief), who dominated the government for two decades; and Sang Hongyang 桑弘羊 (ca. 141–80 B.C.), the son of a merchant family who was appointed by Emperor Wu to one of the three highest positions, imperial counsellor (*yushi dafu* 御史大夫, also translated as censor-in-chief), and who greatly influenced the economic policy of the government. The "reformist" arguments most often were expressed by scholars. Though many of them served in various positions in the bureaucratic system, the scholars claimed their authority not from their political positions but from their cultural capital – knowledge of antiquity, mastery of ancient texts, the ability to read signs from Heaven, and the community of teaching and learning. Representatives of these scholars include Dong Zhongshu (ca. 179–104 B.C.), the most influential Han Confucian scholar, who never held high office in the government; Liu Xiang (79–8 B.C.), an imperial relative whose influence, nevertheless, was mostly as a great scholar supporting reformist policies; and Liu Xiang's son Liu Xin (d. 23 A.D.).

In terms of policy, the modernist statesmen continued the policies of the Qin Dynasty. They were interested in the military and economic strength of the empire. Like the Qin emperor, they used reward and punishment for discipline, and controlled the economy by means of a state monopoly of resources. Their foreign policy was expansionist, for they were eager to launch offensive expeditions to extend their territories. Ideologically, they were not willing to submit to the authority of the divine or of antiquity, but rather rationalized their policies in terms of efficiency and practicality. They sustained the conquest cycle of the Five Phases as the state orthodoxy, and defined the throne as the seat of absolute political power.

The reformist scholars, by contrast, challenged the Qin heritage and its cosmological foundation, and defined imperial sovereignty in terms of morality rather than force. The scholars saw imperial government as an instrument for improving the standard of life of the population. They preferred to leave major economic resources for private enterprise, and to use diplomatic rather than military methods in dealing with foreign relations. They tried to restrain the political and military ambitions of the court by invoking morality, and promoted rule by education and ritual rather than by penal law.

The contest between absolute political power and restraining moral authority existed from the beginning of Han rule. The early emperors and statesmen of Han continued most Qin policies. The founding emperor Gao's reign inherited the symbol of Water as the dynastic symbol, the conquest cycle as orthodoxy, the Qin calendar, and the worship of the five *Di.* Emperor Gao, like the Qin emperor, also ascended the throne because of his military achievements. For almost the entire first century of Han, modernist statesmen's continuation of the Qin policies dominated the court. And it was not until a second century passed that the Qin type of sovereignty, justified by the conquest cycle of *Wuxing* and continued by modernist statesmen, was officially replaced by another – sovereignty justified by a generation cycle and promoted by reformist scholars.

While the founder of the Han, Emperor Gao, inherited most of the Qin practices, he also inherited the tension between political power and moral authority that was embedded in the Qin political and ideological systems. When Lu Jia 陸賈 (fl. ca. 206–180 B.C.), a scholar-official, repeatedly advised Emperor Gao to follow the classical texts of the *Book of Poetry* and the *Book of Documents*, Emperor Gao scolded him, saying: "It [the empire] was what I myself won from riding on the warhorse, and what has it to do with the *Book of Poetry* and the *Book of Documents*!"[35] To challenge the sovereignty grounded in the ruler's achievements on the warhorse and to restrain a ruler who could insult scholars by urinating in their ritual caps,[36] Lu Jia, followed by Jia Yi 賈誼 (201-168 B.C.) and other influential scholars of the time, warned the Han emperors by recalling the fall of the Qin, and contrasted Qin and Han as two opposite types of government.

Both Lu Jia and Jia Yi attributed the Qin's downfall to its rule by force and its lack of morality. Their rationale was that, while seizing power depends on violence, keeping power depends on rule by morality. They thus made the statement that ideal government should be based on ethical principles, while accepting the legitimacy of violence in the transmission of power. To answer Emperor Gao when he asked "why the Qin lost the empire and I gained it," Lu Jia compiled *Xinyu* 新語 (New analects), which challenged the cosmological foundation of Qin rule

35 *SJHK* 97, pp. 15–6; *SJ*, p. 2699.

36 According to *Shiji*, Emperor Gao was said to dislike Confucian scholars. Once he took off the scholars' ritual caps and urinated in them. He often scolded them and could not stand a conversation with them. *SJHK* 97, p. 13; *SJ*, p. 2692.

and laid the foundation for the cosmological discourse that was to be developed throughout the Former Han.

Ontologically, Lu Jia submitted the five conquering powers that supported Qin to a moral cosmology of Heaven, Earth, and Man:[37]

> The *Commentary* says, Heaven gives birth to ten thousands things, with Earth nourishing them, and the sage completing them. Achievement and virtue (*de*) integrate, then the Way (*Dao* 道) and art (*shu* 術) are generated. Therefore it is said, opening up the Sun and the Moon, laying out the stars and constellations, ordering the four seasons, adjusting *Yin* and *Yang*, deploying *qi* 氣 and controlling [human] nature (*xing* 性), and then arranging *Wuxing*.

Here, *Wuxing* is submitted to the greater moral cosmology of Heaven, Earth, and Man; it is only one of many manifestations of ethical values that are embodied in the cosmos. In this cosmology, Heaven is the ultimate moral consciousness and source of order. The human realm no longer consists of violent forces conquering one another, but rather of active moral agents completing the work of Heaven with this inner morality. Achievement must be completed with virtue, because it is the virtues that constitute the fundamental dynamism of the cosmos and human history:[38]

> The *Yang qi* is generated by benevolence (*ren* 仁) and the regulation of *Yin* is descended by righteousness (*yi* 義) . . . the *Spring and Autumn* praises and condemns according to benevolence, and the *Book of Poetry* judges survival and extinction according to righteousness. Heaven and Earth come together with benevolence, the eight diagrams follow one another with righteousness.

According to Lu Jia, it is ethical principles that unite Heaven and Man and compose the single cosmic-social order (*Tian ren he ce* 天人合策). Liu Jia did not reject the conquering cycle of *Wuxing*, but rather submitted it to the greater moral cosmology of Heaven, Earth, and Man.

The ontological theorization of cosmology was aimed at a sociopolitical program, that is, to substitute moral quality for material achievements as the basis of imperial sovereignty, and to substitute moral conduct for conquering force as the basis for transmission of power. Lu Jia first moralized the concepts of *Dao* and *De.* In the Daoist classics *Dao De Jing* and *Zhuangzi*, *Dao* is undivided, immanent, and ineffable, free of moral judgment and intention; and *De* is a person's potency to act according to *Dao*, being free from conscious judgment and desire.[39] Lu Jia revived the Confucian interpretation of these concepts, redefining

37 *Xinyu* 1, p. 1a, SBBY edition. 38 *Xinyu* 1, p. 3b.

39 For the analysis of the *Dao* and *De* in different philosophical traditions, see Schwartz, 1985, pp. 62–3, 192–210; and Graham, 1989, pp. 13, 219–23.

Dao as the highest moral principle of human conduct and *De* as the moral quality to realize such a principle. For Lu Jia, imperial sovereignty based on these moral principles – *Dao* and *De* – could not coexist with that based on martial force and self-interest – *wu* 武 and *li* 利. The moral had to replace the martial: "Only when self-interest is eliminated can *Dao* be illuminated, and only when martial force yields can *De* prevail."[40] The transmission of such sovereignty is explained by the moral superiority of the new sovereigns and the corruption of the old ones. The historical examples are the founder of Shang, named Tang, and the duke of Zhou, who both achieved kingly position by "speaking kind words with their mouths and acting according to the *Dao* of goodness."[41]

Lu Jia's theory pioneered the enterprise of constraining political power with a moral cosmology, an enterprise that engaged Han scholars for generations to come. But his pioneering work had no influence on the practice of the court until decades later, after being enhanced by the work of Jia Yi and Dong Zhongshu.

Jia Yi's criticism of Qin and warning to Han emperors can be crystallized in one phrase: "[Qin] failed to rule by humanity and righteousness, yet the power to attack and the power to maintain what one has won are not the same." In advising Han emperors to avoid repeating Qin's mistake, Jia Yi proposed to distinguish the imperial sovereignty of Han from that of Qin by changing the symbol of the empire from Water to Earth. The change of dynastic symbol would involve a series of ritual and institutional reforms that would reinforce the break with Qin:[42]

> Jia Yi held that from the establishment of Han to the regime of Emperor Wen, twenty-some years had passed. All under Heaven was in peace and harmony, it was necessary to change the calendar and ritual color and robes, establish institutions, define the official titles, promote rites and music. He then planned out the change, using the color yellow and the number five, created official titles, and completely changed the Qin law.

Unfortunately, Emperor Wen did not accept Jia Yi's suggestion. Even Gong Sunchen's 公孫臣 much watered-down version of changing symbols was rejected by the court.[43] When finally put into practice in 104 B.C., late in Emperor Wu's reign, most parts of the reform that Jia

40 *Xinyu* 2, p. 6a. 41 *Xinyu* 2, p. 7b. 42 *SJHK* 84, pp. 21–2; *SJ*, p. 2492.

43 Around 168–166 B.C., Gong Sunchen presented a memorial to Emperor Wen, stating: "formerly the Qin received the Power of Water. Now that Han has succeeded Qin, it ought to be under the patronage of the Power of Earth according to the succession of the beginning and the end [of the Five Powers]. The sign for the Power of Earth is the appearance of a yellow dragon. It is proper to change the calendar and alter the ritual color and garments, using the color yellow." But Chancellor Zhang Cang dismissed this suggestion. See *SJHK* 28, p. 40; *SJ*, p. 1381.

Yi proposed had been eliminated, and only a partial symbolic measure was taken:[44]

In the Summer, the Han changed the calendar so that the year began with the first month, chose yellow as the color of the dynasty, and made new official seals so that they all consisted of five characters.

What Emperor Wu's court put into practice was the least revolutionary part of Jia Yi's proposal – his theory of the imperial symbol. In proposing a break with the Qin type of government, Jia nevertheless used the same logic of a conquest cycle of the Five Powers, with Earth conquering Water. Adopting Earth as the symbol of Han implied an acceptance of Qin as a legitimate regime, and of Qin's cosmological foundation supporting such legitimization – destiny in the conquest cycle of mechanical forces.

Earth as the imperial symbol of Han remained orthodox for the rest of the Former Han Dynasty until changed a century later by Wang Mang, Gongsun Shu 公孫述, and Emperor Guangwu 光武, the three men who competed to take over the Mandate of Heaven from the Former Han at the beginning of the first century. Yet the cosmological foundation of such orthodoxy, the conquest cycle of *Wuxing*, was continually challenged.[45] Even when Emperor Wu adopted the new calendar together with the symbol Earth, the calendar represented a new system that competed with the Five Powers – namely, the *Santong* 三統 (Three Unities) system advocated by Dong Zhongshu, the most influential Confucian scholar of the Han period. This system was composed of a cycle of three symbolic colors, each with its related style of government, calendar, and set of institutions. According to this system, Han should adopt the calendar of Xia, which started the year with the first month. The conquest cycle of the Five Powers prescribed that the dynasty conquering Qin should have the ninth month as the beginning of the year, since Qin used the tenth month and Zhou used the eleventh month. By contrast, the Three Unities system of Dong Zhongshu prescribed that Han should adopt the calendar and governmental style of Xia rather than that of Qin. The rationale for changing the calendar according to the Three Unities system was stated by Ni Kuan, Sima Qian, and others: "Today we should adopt the Xia calendar" because "according to the system of the Three Unities, the later sages should succeed the earlier

44 *SJHK* 28, pp. 84–5; *SJ*, p. 1402.

45 The official change of symbol of Han from Earth to Fire was accomplished by Wang Mang a century later. Wang Mang also replaced the conquest cycle with a generation cycle of *Wuxing* as the orthodoxy for dynastic transmission. See the discussion in the rest of this chapter; and Loewe, 1979a.

Table 4.3. *The Five Powers and the Three Unities compared*

Dynasty	Five Powers	Three Unities
pre-Xia	Earth/Yellow/ninth month	Red/Cultivation/eleventh month
Xia	Wood/Blue/first month	Black/Honesty/first month
Shang	Metal/White/twelfth month	White/Reverence/twelfth month
Zhou	Fire/Red/eleventh month	Red/Cultivation/eleventh month
(Qin)	Water/Black/tenth month	(the great chaos)
Han	Earth/Yellow/ninth month	Black/Honesty/first month

ones, two of which have already preceded us."[46] The Three Unities can be compared to the cycle of Five Powers, as shown in Table 4.3.[47]

In terms of format, the two systems appear to be quite similar, both explaining the succession of dynastic power by cyclical rotation, both using a color and the beginning month of the year to represent each stage of the cycle, and both distinguishing types of government according to these stages. But the Three Unities system did not simply use a cycle of three stages to replace that of five, but rather shifted the foundation of sovereignty from force to morality. Using this new system to explain the transmission of power, Dong developed the principles that Lu Jia had laid out, now providing them with a cosmological and historical system. Most importantly, Dong substituted a shift of the Mandate of Heaven for a conquering force as the ultimate cause of the transmission of dynastic power. In the memorials that he sub-

46 *HSBZ* 21A, p. 25a; *HS*, p. 975.

47 Scholars after Dong greatly elaborated on the Three Unities system. Even in *Chunqiu fanglu*, the text attributed to Dong Zhongshu, we find that the Three Unities system is elaborated into a cycle of twelve stages, when the cycle of three colors is further multiplied by four styles of government. Because *CQFL* – at least Chapter 23, which describes the Three Unities system – was compiled after Dong Zhongshu's death, I do not use it for reconstructing Dong's Three Unities system. Instead, I use a more reliable source – Dong Zhongshu's three memorials recorded in *Hanshu*. According to these three memorials, I reconstruct the system of three styles of government matching the three calendars, disregarding the later elaborations of the four styles and twelve stages. The calendar system found in *CQFL* is included in Table 4.3 since the calendar system of the Three Unities system was already well known by 104 B.C., the year of Dong's death. For the calendar system of the Three Unities system, see Su Yu, *CQFLYZ* 7, pp. 10b–12b. For the authenticity of *CQFL*, especially that of Chapter 23, see *CQFLYZ* 7, pp. 25a, 30a; and 16, p. 16b; Arbuckle, 1995, p. 587; and Queen, 1996, pp. 69–112.

mitted to Emperor Wu, Dong reestablished the supreme authority of Heaven:[48]

The ordinances of Heaven are termed Mandate, the Mandate cannot be put into operation without a man of sagely quality . . . This is why the kingly ruler pays careful attention to receiving the intention of Heaven, so that he may conform with the Mandate.

Since the legitimization of a sovereign must come from the Mandate of Heaven, rather than from conquering forces or the personal power of the monarchy, this legitimization had to be verified by Heaven:[49]

The man who received a kingly position because Heaven gave him great support and made him the king must receive signs that are not produced by human effort but rather arrive by themselves. These signs are the verification of receiving the Mandate. When people all under Heaven submit themselves with one heart to such a king as if they were submitting to their parents, then Heavenly omens arrive responding to such sincerity. The *Book of Documents* says, "white fish came into the boat of the King, fire appeared on top of the King's house and transformed into birds." All these are verifications of receiving the Mandate.

Here the auspicious omens used in the system of Five Powers are reinterpreted and transformed – from signs of destiny in a cycle of conquering forces to signs of Heaven's intention, which restrains and supervises political power. Omens no longer represent different types of sovereignty, but rather verify a single permanent sovereignty that derives from the ultimate source of authority – Heaven. Unlike his predecessors Lu Jia and Jia Yi, who had to accept force and violence in the transformation of power while at the same time advocating rule by morality, Dong Zhongshu eliminates force and violence entirely from the transmission of dynastic power, attributing the dynamism of such transmission to the shifting of the Mandate of Heaven.

Dong Zhongshu thus rejects the implication of Five Powers theory that there are different ways of government, each legitimate: rule by force and rule by virtue. He argues that the *Dao* of Heaven is the only legitimate and permanent way of government. The rotation of the three styles of government, accompanied by changes in the calendar and other institutions, serves only to illuminate the Mandate of Heaven and to rectify the deviations from *Dao* by the preceding dynasty. The eternity of *Dao* and the rectifying function of the Three Unities system are forcefully stated in Dong's memorials:[50]

48 *HSBZ* 56, p. 15a; *HS*, p. 2515; translation adopted with minor alterations from Loewe, 1982, p. 150.

49 *HSBZ* 56, p. 4a–b; *HS*, p. 2500.

50 *HSBZ* 56, p. 16a–b; *HS*, p. 2518–19.

With *Dao* there should be no corruption in ten thousand generations, and corruption is caused by the loss of *Dao.* The *Dao* of the former kings was often deviated from and thus was not realized completely. That is why the government had confusion and malfunction. What needs to be done [by the succeeding king] is only to rectify the deviation and cure the corruption . . . Changing the calendar and altering the ritual garment and color are only to be accordant with the Mandate of Heaven. All the rest should follow the *Dao* that [the ancient sage] Yao practiced, and what is there to be changed!? That is why the king has only the name of reforming the institutions without the reality of altering *Dao.* That Xia promoted Honesty (*zhong*) 忠, Shang promoted Reverence (*jing*) 敬, and Zhou promoted Cultivation (*wen*) 文 is because they each succeeded the former by curing its corruption, and to do so they needed to use the proper style . . . Now that Han has succeeded an age of great chaos, it is necessary to reduce the Cultivation of Zhou and to adopt the Honesty of Xia.

Thus with the Three Unities system and the eternal *Dao* of Heaven, Dong Zhongshu completely eliminated Qin, and the kind of government it represented, from the legitimate dynastic cycles, condemning it as grave decay and chaos resulting from deviation from *Dao.*

The Three Unities system was a new explanation of dynastic transmission that competed with the Five Powers, leaving its mark in Emperor Wu's change of the imperial symbol in 104 B.C., that is, his adoption of the Xia calendar for the Han Dynasty. The combination of symbols from the two competing systems in the new protocol of Han – the Power of Earth and the color yellow from the Five Powers, and the Xia calendar from the Three Unities – reflects the competition of two conceptions of sovereignty and two ways of government in the Han, those of modernist statesmen and reformist scholars. As Emperor Xuan (73–49 B.C.) openly declared: "The Han Dynasty has its own institutions and laws, which are variously [taken from] the ways of the 'hegemons' (*ba*) 霸 and the [ideal] kings (*wang*) 王."[51] In its mixing of the two conflicting theories of sovereignty, Emperor Wu's change of the imperial symbol marks a decline in the modernists' domination of state policy and the growth of the ideological, social, and political force of reformists.

Fire and the Generation Cycle of Wuxing

While Dong Zhongshu challenged the system of Five Powers and the type of sovereignty it legitimated with a rival system of Three Unities, his followers carried the challenge much further by moralizing the *Wuxing*

51 *HSBZ* 9, p. 1b; *HS*, p. 277; translation adopted from Dubs with modifications. See Homer H. Dubs, *The History of the Former Han Dynasty*, vol. 2 (abbreviation *HFHD*), p. 301.

system itself, thus completely transforming the cosmological foundation of imperial sovereignty from within. This moralization was accomplished by substituting a generation cycle of *Wuxing* for the conquest cycle as the foundation of imperial sovereignty, and accordingly changing the dynastic symbol of Han once again, from Earth to Fire.

The vigorous contest between modernist statesmen and reformist scholars was articulated in the competition between these two cycles of *Wuxing*. Under the edict of Emperor Zhao in 81 B.C., the two competing forces openly confronted each other at a conference. On one side were a few high statesmen of the central government, led by the imperial counsellor Sang Hongyang; on the other side were over sixty scholars selected by local officials and representing local interests. In an account of the conference written by Huan Kuan one reign later, entitled *Yantielun* 鹽鐵論 (Discourses on salt and iron), imperial counsellor Sang Hongyang used the conquest cycle of *Wuxing* to justify the modernist policy that he had been implementing:[52]

> The Five Conquering [Powers] succeed one another . . . The four seasons and *Wuxing* repeat their cycles of decline and rise, because *Yin* and *Yang* differ in kind and Water and Fire differ in function . . . Metal rises in the fourth month, giving minor punishment. That is why wheat dies in the summer . . . Spring and summer are the time for birth and growth, and it is beneficial to practice benevolence. Autumn and winter are the time of killing and reserving, and it is beneficial to use punishment.

For a statesman like Sang Hongyang, punishment and military expansion were, like virtue, essential and equal aspects of ruling. And each phase of *Wuxing*, like each of the four seasons, had its own distinctive nature and function that was just as valuable as the others. The phases conquer one another in succession in the same way that the seasons change, and in this fluid cycle there is no single phase, and no type of policy with which it correlates, that is superior to any other.

To oppose the imperial counsellor, the scholars promoted the only justified way of government, ruling by virtue and transforming the people by means of morality. Making this political statement in cosmological terms, the scholars interpreted the succession of *Wuxing* and the four seasons using a generation cycle, substituting birth and nurturing for conquest. They thus rejected any justification for violence: "Weapons [Metal] are instruments of violence. Hard armor and sharp weapons cause the misfortune of the world. Only by using the mother [phase in

52 The conference was recorded later by Huan Kuan during Emperor Xuan's reign in *Yantielun* (Discourses on salt and iron). See Huan Kuan, *Yantielun*, (1954) 1983 (abbreviation *YTLJZ*), p. 570.

Wuxing] to control the son [phase], can it last long." When criticized on the grounds that "practicing virtue in autumn and winter is to go against the *Dao* of Heaven," the scholars redefined the *Dao* of Heaven as a moral consciousness rather than the natural order that justifies violence: "The *Dao* of Heaven favors birth and dislikes killing, favors reward and dislikes punishment . . . devalues winter and values summer."[53] The scholars thus eliminated violence and punishment from the *Dao* of Heaven, submitting the natural order to the moral evaluation of Heaven.

Facing such conflicts between the Three Unities and Five Powers, between the conquest and generation cycles of *Wuxing*, and between Heaven as the highest moral consciousness and *Wuxing* as a mechanical order and historical destiny, generations of scholars made a grave effort to reconcile different cosmological systems into a single all-embracing one, making it the embodiment of ultimate ethical principles, values, and authority. This moralizing enterprise eventually turned the political and intellectual tide of the Han Dynasty. The leading figures in this enterprise, the father and son Liu Xiang and Liu Xin – members of a scholar-official family that was related to the imperial family – proposed replacing the conquest cycle with a generation cycle as the basis for dynastic transmission and accordingly using a new symbol for the Han Dynasty, that of Fire:[54]

> Liu Xiang and his son [Liu Xin] believed that *Di* 帝 originated from *Zhen* 震; that is why Baoxishi 包羲氏 began with the Power of Wood. After that the succession went from the mother to the son to the end of the cycle and started all over again. It went from Shennong 神農 and Huang Di 黃帝 (Yellow Emperor) to the three dynasties of Tang 唐 and Yu 虞, and then Han received Fire. That is why when Emperor Gao was rising, there was a goddess mother crying in the night, showing him the verification of the Red Di. Emperor Gao had his flags and garments all in red, since then [Han] received the Unity of Heaven (*Tian tong*) 天統. Formerly Gong Gong 共工 intruded between Wood and Fire with his Power of Water just like the destiny of Qin, and neither of them could last because they were not in the cycle of succession.

Even though *Wuxing* remains its basis, there are three fundamental changes made to the *Wuxing* system in this proposal of Fire as the new symbol of Han. First, Fire as the symbol of Han is a result of using *Wuxing* to synthesize two other cosmological systems, those of *Yijing* and the Three Unities. Using "*Di* originated from *Zhen*" to support the dynastic transmission in the sequence of the generation cycle of *Wuxing* indicates that *Wuxing* now was combined with the Eight Trigrams of *Yijing*. That Han received the Unity of Heaven and the symbol of Fire resulted from

53 *YTLJZ*, pp. 570–1. 54 *HSBZ* 25, pp. 23b–24a; *HS*, pp. 1270–1.

combining and reforming the Five Powers and Three Unities systems. Therefore, the formerly conflicting systems of cosmology were now synthesized into a unified whole.

Secondly, retaining *Wuxing* as the basis for transmission of dynastic power, Liu Xiang and Liu Xin nevertheless changed the conquest cycle of Five Powers to the cycle of generation from mother to son, thus shifting the ground of imperial sovereignty from conquering force to nurturing morality, again without discarding the *Wuxing* system. While the conquest cycle of *Wuxing* explains the dynastic transmission only in one cycle of five dynasties, the generation cycle of mother-to-son succession extends back for three cycles, including fifteen dynasties.[55]

Thirdly, the two Lius used the generation cycle of *Wuxing* to achieve what Dong's Three Unities system was meant to achieve, that is, the exclusion from legitimate dynastic succession of dynasties that were symbolized by sheer force and violence. The Lius further suggested that the Qin dynasty was not a historical accident; it had an antecedent, Gong Gong, and more importantly, possible successors. Yet all of these dynasties were doomed to be short-lived because they were "not in the cycle of succession."

Finally, the Lius claimed that Heaven was the supreme moral authority and that a legitimate dynasty must receive omens from Heaven to verify having received Heaven's mandate, such as the god mother crying for Emperor Gao.

Liu Xiang and Liu Xin's theory marks the systematic transformation of a mechanical cosmology of conquest into a moral cosmology of generation and nurturing. Even though the generation cycle and the symbol Fire were not adopted by the Han court during Liu Xiang's time, Liu Xin lived to assist Wang Mang in establishing the New Dynasty (Xin, 9–23 A.D.), establishing the generation cycle of *Wuxing* as the orthodoxy of dynastic succession, with Fire as the symbol of Han and Earth as the symbol of the succeeding New Dynasty of Wang Mang. After Emperor Guang Wu (ruled 25–57 A.D.) took over Wang Mang's power and restored the Han (25–220 A.D.), he adopted the generation cycle of dynastic succession and Fire as the symbol of Han, both of which became the state orthodoxy throughout the Later Han.[56]

55 Liu Xin used the generation cycle of *Wuxing* to account in detail for dynastic succession since the beginning of human history, thus systematically reconstructing the history of antiquity. His reconstruction became the most orthodox and was adopted by official histories thereafter. Liu Xin's system of dynastic succession is recorded in the "Treatise of Pitchpipes and Calendars" of *Hanshu*. See HSBZ 21B, pp. 45a–76b; *HS*, pp. 1011–24.

56 For Wang Mang's changing dynastic symbol, see the later part of this chapter. For a full account of changes in the dynastic symbol, including the failed effort of Gongsun

It was this theory that was adopted by Ban Gu in "*Wuxing* zhi," which replaced "Hongfan"'s cycle of *Wuxing* with the generation cycle, representing the first change to the structure of "Hongfan." While this structural change was the first achievement of the moralization of cosmology, the second structural change to "Hongfan," made by the compiler and authors of "Wuxing zhi," represents the complete transformation of *Wuxing* from a cosmology that justifies conquest and violence to a moralized, all-embracing cosmology.

Earth, Man, and Heaven

In addition to substituting a generation cycle for the conquest cycle of *Wuxing*, the compiler and authors of "Wuxing zhi" made a second structural change to the canon of "Hongfan," that is, the change from nine categories to the Earth, Man, and Heaven cosmology of "Wuxing zhi."

While based on the canon of "Hongfan" and using *Wuxing* in its title, "Wuxing zhi" is structured neither by the nine categories of Hongfan nor by the five categories of *Wuxing*. Instead, the compiler and authors of "Wuxing zhi" used the nine categories of "Hongfan" to reconstruct a new cosmology of Earth, Man, and Heaven. The introduction to "Wuxing zhi" quotes "Hongfan," showing that "Hongfan" was structured in nine categories, with *Wuxing* as one of them. The structure of the quoted canon is based on the nine categories listed in numerical sequence: "The first one is called X, the second is called Y . . ." This can be represented as follows:[57]

1. *Wuxing* 五行
2. The Five Duties (*wu shi*) 五事
3. The Eight Policies (*ba zheng*) 八政
4. The Five Regulators (*wu ji*) 五紀
5. Sovereign Standard (*huang ji*) 皇極
6. The Three Powers (*san de*) 三德
7. The Examination of Doubts (*ji yi*) 稽疑
8. Several Verifications (*shu zheng*) 庶徵
9. The Five Blessings and the Six Extremities (*wu fu, liu ji*) 五福, 六極

"Hongfan" describes these nine categories as resources that a ruler uses to maximize his power. According to Michael Nylan, the original "Hongfan," dated to the late fourth century B.C., was a ruler's manual

Shu and the policy of Emperor Kuangwu, which this study does not address, see Loewe, 1979a; and *HHSJJ* 1A, p. 18b; *HHS*, p. 27; *HHSJJ* 13, p. 16b; *HHS*, p. 538.

57 *HSBZ* 27A, pp. 2b–3a, *HS*, p. 1316.

on administration during the Warring States period; it was then transformed into a sacred canon on cosmology during the Han.[58] The compiler and authors of "Wuxing zhi" selected only categories 1, 2, 5, and 8 of this canon and reworked them. They used category 1 – *Wuxing* – to construct the realm of Earth, categories 2 and 8 – the Five Duties and Several Verifications – to construct the realm of Man, and category 5 – the Sovereign Standard – to represent the realm of Heaven. In this way, the compiler and authors of "Wuxing zhi" used "Hongfan" categories to reconstruct a moral cosmology of Earth, Man, and Heaven.

After such structural rearrangement, "Wuxing zhi" served to reinterpret the meaning of each of the "Hongfan" categories selected, changing these categories from the ruler's political resources to a moral cosmology of resonance between Heaven and Man. This change was further completed by layers of interpretation.

The realms of Earth, Man, and Heaven are each introduced by a quotation from "Hongfan," followed by an interpretation of the quotation from the *Commentary* on *Wuxing* by Fu Sheng 伏生 (fl. ca. 221–170 B.C.), a scholar of the Qin and early Han.[59] The *Commentary* is then further interpreted by Han scholars such as the Ouyangs 歐陽,[60] Xiahou Shichang 夏侯始昌 (fl. 96 B.C), and Xiahou Sheng 夏侯勝 (fl. 74 B.C.),[61] whose texts were officially listed among those to be studied by the academicians.[62] Finally, all these interpretive theories were exemplified by cases of bad omens found in the *Spring and Autumn* and in the Han Dynasty, interpreted by Dong Zhongshu, Liu Xiang, Liu Xin, Jing Fang,

58 Nylan, 1992, pp. 13–44, 105–48.

59 The full title of Fu Sheng's commentary is *Shangshu dazhuan* 尚書大傳 (Great commentary on the *Documents*); it was submitted to the Han court between 179 and 157 B.C. See *Shangshu dazhuan*, SBCK ed.

60 *Hanshu*'s "Biography of Scholars" states that the Ouyangs have been an influential scholar family throughout the Han. Ouyang Sheng 歐陽生 first studied with Fu Sheng, and transmitted his learning of the *Documents* to Ni Kuan 倪寬, who discussed the *Documents* with Emperor Wu (140–87 B.C.). Ouyang Sheng's son then studied with Ni Kuan, and the tradition was transmitted across generations to the great grandson, Ouyang Gao 歐陽高, who was an academician. Thus the theory of the Ouyangs came from the teaching of Ni Kuan. The descendants of the Ouyangs became academicians, participated in the Shiqu 石渠 conference (51 B.C.) in Emperor Zhao's time, and held academic positions during Wang Mang's reign. See *HSBZ* 88, pp. 11b–12a; *HS*, p. 3603.

61 The theories of Xiahou Shichang and Sheng also came from the tradition of Fu Sheng and Ni Kuan. Xiahou Shichang was a member of the senior generation of the Xiahou family and an eminent academician during Emperor Wu's reign. Xiahou Sheng, a member of the Xiahou clan, studied the *Documents* from Shichang and became famous for his interpretation of bad omens according to the *Documents*. He taught emperor dowagers and served as the grand tutor for the heir apparent during Emperor Xuan's reign (73–49 B.C.). See *HSBZ* 75, pp. 2a–5b; *HS*, pp. 3155–9.

62 The identification of the sources of these quoted interpretations is based on the commentary by Wang Mingsheng, in *HSBZ* 27A, p. 2b.

and Ban Gu. Each layer of the interpretation represents a further step in building correlation and resonance among the realms of Heaven, Earth, and Man.

The Realm of Earth

The realm of Earth starts with a quotation from "Hongfan" stating the sequence of *Wuxing* and the properties of the five *xing*, in the following formula:

The first [in the sequence of *Wuxing*] is called (*yue* 曰) X, . . .
X is called to be able to Y

The quotation can be illustrated by the following list:[63]

1. Water – to soak and descend
2. Fire – to blaze and ascend
3. Wood – to be curved and straightened
4. Metal – to conform and change
5. Earth – to sow and harvest

This quotation from "Hongfan" simply describes *Wuxing* as five basic materials, each having its physical property and function. The connection between *Wuxing* and government is not clearly stated. *Wuxing* here represents the five kinds of material resources that the ruler regulates to provide the economic basis for the government.[64]

But Fu Sheng's commentary starts to establish clear correlation between *Wuxing* and government. It attributes the loss of physical properties of these five kinds of material to malfunctions in the government, thus using *Wuxing* as a correlative schema connecting human behavior with cosmic order. The correlation thus established by Fu Sheng is shown in five lists of malfunctions in the government. Each malfunction causes a loss of the property associated with one of the Five Phases:[65]

The Phase of Wood:
Hunting out of season;
Failure to present ritual offerings of food and drinks;
Unregulated exiting and entering;
Depriving people of the time for farming;
The existence of treacherous plots:
these cause Wood to lose its property of being crooked or straight.

63 Translation adopted, with modifications, from Beck, 1990, pp. 132–3.
64 Nylan, 1992, p. 15.
65 *HSBZ* 27A, pp. 3a, 4a–b, 16b, 17b, 19a; *HS*, pp. 1318, 1320, 1338, 1339, 1342.

The Phase of Fire:
Disregarding the laws;
Expelling meritorious ministers;
Killing the heir apparent;
Taking a concubine as the principle wife:
these cause Fire to fail to blaze and ascend.

The Phase of Earth:
Constructing palaces and terraces;
Sexual licentiousness within;
Transgressing the proper relations between relatives;
Insulting the father or elder brothers:
these cause failure in sowing and harvesting [because Earth loses its property].

The Phase of Metal:
Indulging in offensive warfare;
Ignoring the well-being of the people;
Transgressing the borders:
these cause Metal to fail to conform and change.

The Phase of Water:
Negligence in attending to ancestral shrines;
Failing to pray at the altars;
Abandoning ritual sacrifices;
Going against the temporal order of Heaven:
these cause Water not to soak and descend.

Fu Sheng's commentary thus transforms "Hongfan"'s *Wuxing* from five kinds of material to five phases of correlation that verify functions and malfunctions of the government, with Wood correlated mainly to food production, Fire to law and the political system, Earth to internal affairs of the court, Metal to military actions and foreign relations, and Water to ritual performance. Yet Fu Sheng's correlation is still very preliminary; it states the direct correspondence of the Five Phases to five kinds of governmental function, but with neither elaborate cosmological correlation nor a theory of the resonance provided.

However, the theories of the Ouyangs and the Xiahous, which are quoted after each quotation of "Hongfan" and Fu Sheng's *Commentary*, elaborate on Fu Sheng's interpretations in three respects. First, they extend the correlation of the Five Phases to include directions and the system of *Yijing*. Second, they focus on the sovereign's personal behavior as the point of criticism, rather than on a general failure of governmental functioning. And third, they list specific kinds of bad omens as verifications of the sovereign's bad behavior, which in turn causes disorders in each of the Five Phases. For example, for the Phase of Wood, they say:[66]

> **In the Phase of Wood,**
> **The direction** (*fang*) is East,
> **In** ***Yijing***, it is "*Guan*" – the Wood on Earth,
> **In the king's affairs**, it is manner and demeanor that are visible. That is why the king has to regulate his steps of walking by wearing jade instruments, to control his speed of riding according to the metal bells on the carriage, to follow the regulation of the three types of cooking during hunting, and to observe rituals of offering food and drinks. His exiting and entering shall have proper names, his uses of people shall be at the proper time. His duty is to promote agriculture and sericulture, and his concern shall be the stability of the people – **only then will Wood preserve its property.**
> **If** he hunts extensively and does not return to the palace, indulges in eating and drinking ignoring regulations, robs people of time by excessive uses of tax and services and making treacherous plots to destroy people's wealth – **then Wood will loose its property.**
> **When Wood looses its property of being crooked or straight**, the carpenters often fail in making wheels [by crooking the wood] or making arrows [by straightening wood], and trees have monstrous transformations.

Table 4.4 illustrates the complete system of the Ouyangs and the Xiahous, which established elaborate correlations between the Five Phases and the sovereign's behavior.[67]

Following these layers of theoretical interpretation of "Hongfan"'s *Wuxing*, "Wuxing zhi" presents forty-five cases of bad omens, chronologically arranged from the Spring and Autumn period (722–481 B.C.) to the end of the Former Han Dynasty (8 B.C.). These bad omens are further interpreted by Dong Zhongshu, Liu Xiang, Liu Xin, and Ban Gu. The first part of "Wuxing zhi" constructs the realm of Earth, which prescribes the proper behavior of the sovereign using the five earthly phases. By layers of interpretation, *Wuxing* is transformed from five material resources, as seen in "Hongfan," to correlations between cosmic phases and the sovereign's behavior. The theories of such correlation are verified with abundant cases from the past.

66 *HSBZ* 27A, p. 3a–b; *HS*, pp. 1318–9; the author's own emphasis is intended to show the structural format of the argument.

67 Table 4.4 is based on *HSBZ* 27A, pp. 3a–b, 4a–b, 16b, 17b, 19a–b; *HS*, pp. 1318–9, 1320, 1338–9, 1342.

Table 4.4. *Correlations of the Five Phases in the theories of the Ouyangs and the Xiahous*

Phases	Wood	Fire	Earth	Metal	Water
Direction	East	South	Center	West	North
King's affairs	Demeanor	Enlightened governing	Internal affairs	Military affairs	Rituals
Good government	Promoting farming	Order and hierarchy	Moral education	Restraint	Observance and reverence
Bad government	Indulging in hunting, depriving people of the time for farming	Evil conquers upright	Extravagance, licentiousness	Greedy, despising human life	Not reverent, giving untimely orders
Bad omens	Failure in making wheel and arrow, etc.	Fire at palaces and temples	Failure in harvest	Failure in melting and casting metal	Flood, constant rain

Table 4.5. *Correlations of the Five Duties: the merger of "Hongfan"'s categories 2 and 8*

Duty	Demeanor	Speech	Sight	Hearing	Thought
Property	Respect	Compliance	Clarity	Perceptiveness	Capaciousness
Result	Gravity	Orderliness	Wisdom	Deliberation	Sageliness
Verification	Rain	Sunshine	Heat	Cold	Wind
Good result	Gravity	Orderliness	Wisdom	Deliberation	Sageliness
Verified by good omen	Seasonable rain	Seasonable sunshine	Seasonable heat	Seasonable cold	Seasonable wind
Bad result	Wildness	Assumption	Indolence	Haste	Stupidity
Verified by bad omen	Constant rain	Constant sunshine	Constant heat	Constant cold	Constant wind

The Realm of Man

The second part of "Wuxing zhi," which is titled "Five Duties" (*Wu shi* 五事), constructs the realm of Man. The central section of the text, this part is the most elaborate in theory and extensive in content. Of 378 cases of bad omens collected in "Wuxing zhi," 190 are classified in the realm of Man.

Stylistically consistent with the realm of Earth, the section on the "Five Duties" of Man starts with a quotation from "Hongfan." But this time, the quotation is a combination of two separate categories of "Hongfan": category 2, the "Five Duties," and category 8, "Verifications." It first lists the "Five Duties": (1) demeanor, (2) speech, (3) sight, (4) hearing, and (5) thought. It then attributes to each duty its property and the good and bad results of this property, repeating the format of the *Wuxing* section:

The first [in the sequence] is called X (demeanors) . . .
X (demeanors) is called to be able to Y (respect) . . .
Y (respect) makes for (*zuo* 作) Z (gravity) . . .

Immediately after the "Five Duties" comes category 8 – "verifications" – correlating the Five Duties with favorable and unfavorable omens, verifying success and failure in the Five Duties. "Wuxing zhi" therefore combines categories 2 and 8 of "Hongfan" into a seamless coherent correlative schema, as represented in Table 4.5.[68]

68 Table 4.5 is based on quotations from "Hongfan" in *HSBZ* 27Ba, pp. 1a–2a; *HS*, p. 1351.

By combining categories 2 and 8 of "Hongfan" into a coherent correlative system and excluding other categories, "Wuxing zhi" highlights its central theme – correspondence between the conduct of the sovereign and the phenomena of the universe that verify proper or transgressive conduct. Following this quotation from "Hongfan," "Wuxing zhi" then presents Fu Sheng's *Commentary*, which further lists the specific kinds of bad omens induced by the bad conduct of the sovereign in each of the Five Duties. For the lack of *respect* in the Duty of *Demeanors*, it says:[69]

Its **guilt** is *wildness*,
Its **punishment** is *constant rain*,
Its **extreme** is *evil*.
Sometimes there are **goblins** (*yao* 妖) in the form of *clothes*.
Sometimes there are **plagues** (*nie* 孽) of *tortoises*.
Sometimes there are **disasters** (*huo* 禍) involving *chickens*.
Sometimes there are **illnesses** (*e* 痾) of *lower parts of the human body growing upward*.
Sometimes there are **internal omens** (*sheng* 眚) and **external omens** (*xiang* 祥) in the color of *blue*.
These mean the Phase of *Metal* is **harming** (*li* 沴) the Phase of *Wood*.

Following this format, Fu Sheng's system of bad omens can be represented in Table 4.6.[70]

Fu Sheng's interpretation gives "Hongfan"'s categories new meaning. First, it introduces the concept of "punishment." In "Hongfan," both good and bad omens are verifications of the conduct of the sovereign; they are represented as matters of fact, with no further theory about the moral intentions behind them. The quotation from Fu Sheng's *Commentary*, however, omits good omens altogether and elaborates only on bad omens. By presenting bad omens as "punishments," it implies a moral intention behind all the omens, and transforms factual verifications into messages from a moral authority. Secondly, Fu Sheng extends the range of omens from meteorological phenomena, such as rain and sunshine, to include all phenomena in the universe – the world of human beings, animals, plants, and spirits. He thus replaces the causal relation between a single kind of conduct and a single meteorological phenomenon with a comprehensive correlative cosmology, in which the conduct of the sovereign can induce serious disorder in all realms of the universe – in Heaven, Man, and Earth. Thirdly, Fu Sheng uses

69 *HSBZ* 27Ba, p. 2a–b; *HS*, p. 1352.
70 Table 4.6 is based on quotations from Fu Sheng's commentary in *HSBZ* 27Ba, pp. 2a–b, 16a; *HSBZ* 27Bb, pp. 1a, 11a–b; *HSBZ* 27Ca, p. 1a; *HS*, pp. 1352, 1376, 1405, 1421, 1441.

Table 4.6. *Correlations of the Five Duties in Fu Sheng's commentary*

Duty	Demeanor	Speech	Sight	Hearing	Thought
Guilt	wildness	assumption	indolence	haste	foolishness
Punishment	constant rain	constant sunshine	constant heat	constant cold	constant wind
Extremity	evil	sorrow	sickness	poverty	shortening the life
Goblin (*yao*)	in clothes	in poetry	in grass	in drums	in lard and night
Plague (*nie*)	of tortoises	of shell insects	of worms	of fish	of flowers
Disaster (*huo*)	in chickens	in dogs	in sheep	in pigs	in oxen
Illness (*e*)	lower parts of body growing upward	in mouth and tongue	in eyes	in ears	in heart and belly
Internal omen (*sheng*)	in blue	in white	in red	in black	in yellow
External omen (*xiang*)	in blue	in white	in red	in black	in yellow
The Five Phases	Metal harms Wood	Wood harms Metal	Water harms Fire	Fire harms Water	Metal, Wood, Water, and Fire harm Earth

the Five Phases to explain the bad omens that appear in the realm of Man. The separate categories of "Hongfan," categories 1, 2 and 8, are now integrated into a larger correlative cosmology.

The Ouyangs, the Xiahous, and other Han scholars further elaborated upon Fu Sheng's *Commentary*. While in both "Hongfan" and the *Commentary* the agent who induces the bad omens is only implied, various Han theorists explicitly identify this agent as the single human sovereign. For example, for the Duty of Speech, the Han theorists say, "If the **emperor** (*shang* 上) gives orders that are opposing people's hearts, pretentious and disorienting, then he cannot govern the empire." Similarly, for failure in the Duty of Sight, "It means the **emperor** has lost clarity . . . and failed to distinguish good from evil." For the failure in the Duty of Hearing, "It means the **emperor** listened with bias and therefore resulted in haste." And for the failure in the Duty of Thought, "It means the **emperor** is not broad-minded and cannot bear with his subjects, and therefore should not preside over the throne." Explicitly identifying the emperor as the sole agent responsible for the bad omens, the Han theorists used "verifications," which in "Hongfan" were meant to be "objective" measurements of the ruler's and the government's administrative functioning, as vehicles for constructing norms for the moral behavior of the emperor.

Defining the norms for the emperor, these theorists use the concept of *qi*, the fluid cosmic energy, to explain the correlation between the sovereign's conduct and omens:[71]

> If the sovereign (*jun* 君) has excessive *Yang* as to become a tyrant, then subjects will be silenced by fear of punishment, the *qi* of discontent will be expressed in songs and poems, that is why there are goblins (*yao* 妖) in forms of poetry. The plague of shell insects, the small ones with shells and flying around, is also generated by the *qi* of *Yang*.

Bad omens, according to such theorists, are produced by variations in the emperor's *qi*. Omens thus produced are no longer a measurement of the function of the administration, but "punishment" for the emperor. These theorists further develop Fu Sheng's idea of "punishment" by combining it with the concept of *qi*: "When the four *qi* (of demeanor, speech, sight, and hearing) are all in disorder, the punishment is constant rain . . . and its extremity is shortening of lives."[72] The same *qi* is also used to explain the 'harm' among the Five Phases: "When *qi* [of different phases] damage one another, it is called 'harm'

71 *HSBZ* 27Ba, pp. 16a–17b; *HS*, pp. 1376–7. 72 *HSBZ* 27Ca, pp. 1a–b; *HS*, p. 1441.

(*li* 沴)."[73] With the concept of *qi*, therefore, these Han theorists moved one step further from the implicit causal relation between government and verifications toward an explicit cosmology of resonance through the universal energy of *qi*.

For these Han theorists, bad omens were a means of prescribing the norm for emperorship, and of discipline and punishment of the emperor's transgression of the norms. By categorizing the bad omens into an elaborate classification system, the Han theorists could measure the degree of the sovereign's transgression:[74]

In the world of plants, the omens are called "*yao* 妖" (goblins). "*Yao*" is like "*yao tai* 夭胎" (the early fetus), meaning [the problem] is still small. In the world of insects, the omens are called "*nie* 孽" (plague). "*Nie*" is like "*ya nie* 牙孽" (sprouts). Yet when the omen involves the animals, it is called "*huo* 禍" (disasters), meaning the problem is getting bigger. When the omen involves human beings, it is called "*e* 痾," which is the illness. The illness (*e*) means the problem is pervasive and deep. When [the problem] gets even more serious, strange beings will be generated, which are called "*sheng* 眚." If the strange beings are generated externally, they are then called "*xiang* 祥" . . .

Therefore, "Hongfan"'s simple meteorological phenomena used as checks on administration were first expanded by Fu Sheng into categories of omens in the universe, then further developed by Han theorists into a measure of degrees of the emperor's transgression and punishment.

The Realm of Heaven

The part of "Wuxing zhi" devoted to Heaven does not open with a quotation from "Hongfan"; instead, it goes directly to Fu Sheng's *Commentary*. Here the *Commentary* reinterprets category 5 of "Hongfan," the Sovereign Standard. The Sovereign Standard is the heart of "Hongfan"; it states the principles and techniques that the ruler must use in order to forge societal consensus and thus maximize his power.[75] But Fu Sheng uses this term to represent another realm, the realm of Heaven, where the sovereign is being judged, controlled, and punished. He treats the Sovereign Standard as an extension of the Five Duties, giving it exactly the same format. For the sovereign's failure to establish the standard, he says:[76]

73 *HSBZ* 27Ba, p. 2b; *HS*, p. 1354. According to *Zhongwen dazidian*, the term *li* has various meanings, such as blockage of flowing water, the disharmony of different kinds of *qi*, and disorder or chaos.

74 *HSBZ* 27Ba, pp. 2a–2b; HS, p. 1353.

75 Nylan, 1992, pp. 23–32.

76 *HSBZ* 27Ca, p. 10b; HS, p. 1458.

Its **guilt** is *blindness*,
Its **punishment** is *constant yin*,
Its **extreme** is *weakness*.
Sometimes there are goblins (*yao*) in the form of *shooting*.
Sometimes there are plagues (*nie*) of *dragons and snakes*.
Sometimes there are disasters (*huo*) involving *horses*.
Sometimes there are illnesses (*e*) of *inferiors attacking the superiors*.
Sometimes there are *irregularities in the paths of the sun and the moon, and the retrograde movements of planets and constellations*.

Following Fu Sheng's statement are the various Han theories that further interpret it. In "Hongfan" as well as in Fu Sheng's *Commentary*, the title of category 5, "*Huang ji*," is implicit and ambiguous, open to multiple interpretations and difficult to translate.[77] Once interpreted by the Han theorists, it became explicit, focusing on the sovereign and his responsibility. The Han theorists fixed the ruler, his bodily conduct and inner morality, as the responsible agent for both human and cosmic order and disorder:[78]

"*Huang* 皇" is the ruler. "*Ji* 極" is the Center . . . When the ruler fails in all the Five Duties – demeanor, speech, sight, hearing, and thought – he loses his presidency over the Center, and therefore cannot establish ten thousand things.

The theorists thus identify the sovereign as the only human figure responsible for realizing Heaven's ultimate order in the universe, saying: "The king is the one from below who receives from Heaven and orders all things." Whereas failures in the Five Duties induce different degrees of omens as punishment, failure in the Sovereign Standard means the ultimate loss of Heaven's Mandate. When the ruler fails this ultimate responsibility of the sovereign, "The *qi* of Heaven is disturbed, and that is why the punishment is constant *Yin* . . . And the *Yin qi* moves, thus producing the plagues of dragons and snakes."[79] Punishment is the execution of the emperor, ordered by Heaven and put in the hands of either a new sovereign or a usurper:[80]

When the ruler is confused and weak, he becomes the one whom people rebel against and Heaven means to eliminate. If he is not executed by an enlightened sovereign, he suffers the disaster of being assassinated by a usurper. That is why [the *Commentary*] says that there is illness of inferiors attacking superiors.

In the realm of Heaven, the role of the Five Phases seems to have "vanished completely" and the connection between the Five Phases and the

77 It can be translated as "greatness maximized" or "sovereign standard"; see Nylan, 1992, p. 23.
78 *HSBZ* 27Ca, pp. 10b–11; *HS*, pp. 1458–9. 79 Ibid. 80 Ibid.

omens seems to be "increasingly loose," as Mansvelt Beck has noticed.[81] Yet this is not a structural defect, but rather a constructive device for moralizing *Wuxing* and integrating it into a grand moral cosmology of Heaven. The diminishing role of *Wuxing* was necessary in order to elevate Heaven as the ultimate moral authority. After all, for the Han scholars, *Wuxing* as embodied in the earthly Five Phases and the Five Duties of human beings served only to verify the moral order of Heaven. And Heaven, as the supreme moral and divine authority, was not to be equated with the realms of Earth and Man. The Han theorists quoted here explicitly explain why the terms of conquest and harm among the Five Phases shall not be applied in the realm of Heaven:[82]

> The reason why [Fu Sheng] did not say 'the Five Phases harms Heaven' . . . is because it is as if the inferiors dare not harm Heaven. This is similar to [the author of] the *Spring and Autumn*, who says "the king's troops were defeated at the place of *maoyong*," without mentioning who defeated them. The intention [in both cases] is to revere the superior.

Therefore, the compiler and authors of "Wuxing zhi" abandon the consistency of the Five Phases in the last part of the text for the sake of a greater cosmological coherence, that is, the completion of the moral cosmology of Earth, Man, and Heaven, and the elevation of Heaven above all cosmological systems, such as *Wuxing* and *Yin-Yang*, to become the ultimate authority of divinity and morality.

Conclusion

Some final words about the authorship of "Wuxing zhi" are in order. As mentioned at the beginning of the chapter, Ban Gu was the compiler responsible for the existence of the text. But Ban Gu's contribution was not in creating the cosmological structure of the text, but rather in painstakingly synthesizing the complex cosmological discourse of the Former Han and representing its transformation through the two centuries of the Former Han. The moralization and transformation of the cosmology was accomplished by generations of scholars – Fu Sheng, Lu Jia, the Ouyangs, the Xiahous, Dong Zhongshu, Jing Fang – and then completed by Liu Xiang and Liu Xin. Ban Gu did not sacrifice the complexity of the historical transformation of the cosmology for the sake of textual coherence, but rather honestly represented this complexity by recording most of the controversies within the cosmological discourse.

81 Mansvelt Beck observed this structural character of "Wuxing Zhi" but did not give any explanation. Beck, 1990, pp. 139, 141.

82 *HSBZ* 27Ca, pp. 10b–11; *HS*, pp. 1458–9.

The cosmological structure of "Wuxing zhi" was the product of a long process of transforming a cosmology of the conquering Five Powers that justified violence into a moralized cosmology of Heaven. In the moralized cosmology, the Five Phases were subject to the supreme authority of Heaven and served as the verification and embodiment of Heaven's moral intention.

This moralized cosmology challenged, and eventually replaced, the orthodoxy of the Qin and early Han Empires. Its success toward the end of the Former Han not only signifies the growing force of the reformist scholars in society and government, but also indicates their increased control of the core political institution, emperorship. By submitting the throne to the moral authority of Heaven, antiquity, and cosmology, the reformist scholars finally transformed emperorship from a conquering political power into a vehicle for their ethical ideals. While the founding emperor was a successful conqueror who had established the empire on the warhorse and relied on modernist statesmen for governing, the fourteenth emperor, the last of the Former Han, was enthroned at the age of two by the most powerful scholar-official of the time, Wang Mang.

We can say that the textual structure of "Wuxing zhi" reflects the total transformation of both cosmology and imperial sovereignty during the Former Han. This transformation parallels the political practice that ended the Former Han Dynasty. Wang Mang was the greatest practitioner of the Han Confucian scholars' ideals and theories. Even from Ban Gu's unfavorably biased biography of him, one can discern that Wang Mang was not only a true scholar, but also spent his life realizing Confucian ethical ideals and restaging the Confucian golden age of antiquity in real life. He ended the Former Han by taking the throne himself and initiating the New (Xin 新) Dynasty. He succeeded in ending the Former Han without a civil war, through "abdication" rather than military conquest. His success was not achieved by his "hypocritical" and "treacherous" plots, as portrayed in the biography in *Hanshu*, but was based on the ideological and political preparation of generations of Han scholars.

Wang Mang's enthroning edict illustrates how he put the Han scholars' theory into practice. In January of 9 A.D., Wang Mang announced his final step toward taking the throne of emperorship:[83]

> I possess no virtue, [but] I rely upon [the fact that] I am a descendant of my august deceased original ancestor, the Yellow Emperor, and a distant descendant

83 *HSBZ* 99A, pp. 35b–36a; *HS* 4095–6; English translation adopted with modificaion from Dubs, *HFHD*, vol. 3, pp. 255–6.

of my august deceased first ancestor, Emperor Yu [Shun], and the least of the Grand Empress Dowager's relatives. August Heaven and the High God have made abundantly apparent their great assistance, so that the mandate [of Heaven] has been completed and the succession [of the imperial rule] has been set in order. By omens and credentials, designs and writings, a metal casket and a written charter, the gods have proclaimed that they entrust me with the myriad common people of the empire.

The Red Emperor is the genius of Emperor Gao of the Han Dynasty. He has received a Mandate from Heaven and has transmitted the throne [to me by] a writing on a metal charter. I have been extremely reverent and awed – [how could I] presume not to receive it respectfully? On the day of *wuchen* 戊辰, which is the day for founding, I wear the royal hat and ascend the throne as the actual Son of Heaven. It is fixed that the title [of my dynasty] in possessing the empire shall be Xin.

In this edict, Wang Mang lists three justifications for his enthronement. First, he claims to be the descendent of the Yellow Emperor representing the Phase of Earth, and according to the birth cycle, Earth is the son of Fire – the Han Dynasty founded by the Red Emperor (Emperor Gao). Second, Heaven has shifted its mandate to Wang Mang by choosing Earth as the current phase. For this, Heaven has sent down numerous auspicious omens as verification. Third, there has appeared a written decree of the Red Emperor, the diseased founding emperor of Han, announcing his abdication of the throne of the Han Dynasty.

Comparing Wang Mang's enthroning edict with that of the First Emperor of Qin, one sees a complete shift in the basis of imperial sovereignty. The Qin emperor was enthroned by his military achievement; Wang Mang was enthroned by Heaven's Mandate. This comparison illustrates the transformation of the ruler from the conqueror of the world to the highest moral exemplar of all human beings, to be imitated by his subjects. By showing himself as a recipient of the Mandate of Heaven, a superhuman authority, Wang Mang submitted political power to moral authority. By the same token, he claimed himself as the highest human representation of this authority, the only man entitled to conduct the highest rituals of communication with Heaven.

This transformation of imperial sovereignty was achieved, at least partially, through the scholars' enterprise of moralizing cosmology, which provided the scholars with the highest moral authority and source of symbolic power. This source of power and authority was actively cultivated and mobilized by political actors, such as Wang Mang and the compiler and authors of "Wuxing zhi." Figure 4.1 shows one of the great moral achievements of Wang Mang that led him to the throne. He proposed to reconstruct the ritual hall of antiquity, called *Bi yong* 辟雍 or

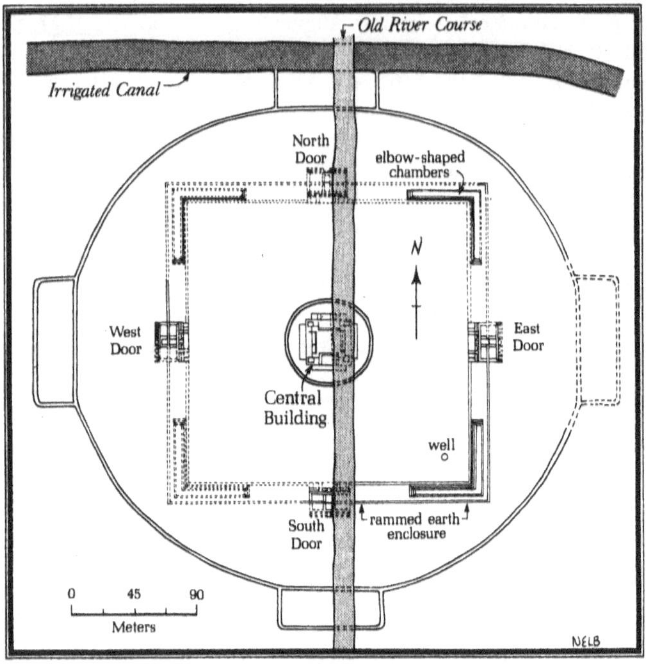

Figure 4.1. Plan of the site of the ritual complex built by Wang Mang as a reconstruction of "*Bi yong*" or "*Ming tang*" of antiquity (Wang Zhongshu, *Han Civilization*, New Haven: Yale University Press, 1982, Fig. 30, p. 25).

Ming tang 明堂, which was believed the highest temple of Shang and Zhou, abolished for a thousand years. With a Bronze Age temple in mind, Wang Mang built the ritual complex that has now been discovered by archaeologists. In the center of the compound was a hall composed of a central room and four side chambers, modeled upon the *ya*-shaped ancestral temple of great antiquity that the Han scholars reconstructed in ritual texts. While repeating the *ya*-shaped *Sifang* structure seen in Shang royal tombs, Wang Mang's ritual hall was located in a much more complicated symbolic context – the *ya*-shaped ritual hall was built on a round platform (60 meters in diameter), which was in turn placed in a large square courtyard composed of four walls (each 235 meters long). Surrounding the square courtyard was again a large circle (360 meters in diameter) formed by a ditch of water.[84]

84 For the excavation report, see Tang Jinyu, 1959, pp. 45–54. For the reconstruction of the ritual complex, see Wang Zhongshu, 1982, figure 30.

The cosmological meaning of this ritual complex echoes that of the textual structure of "Wuxing zhi" in two ways. First, both "Wuxing zhi" and Wang Mang's ritual complex represent a moralized *Wuxing*, with a generation cycle of birth and nurturing replacing the conquest cycle of force and violence. The compiler and authors of "Wuxing zhi" used the generation cycle of *Wuxing* to structure their text; Wang Mang used the generation cycle of *Wuxing* as the pattern for the ritual performance to be carried out within the ritual hall. The central hall and four side chambers represent the Five Phases (*Wuxing*), with each side chamber symbolizing one phase and one season, and the central room symbolizing the Phase of Earth and the middle of the year. Each season, Wang Mang would dress in the proper color and perform the proper ritual in the proper chamber. The movements of the emperor, Wang Mang, acted out the generation cycle of *Wuxing* in a temporal sequence. Therefore, *Wuxing* no longer symbolized the conquering forces of the cosmos, but the moral intention of birth and nurturing of Heaven. Even though the ritual hall repeated the pattern of the *ya*-shaped Shang royal tombs (see Chapter 2), its occupant was now the emperor rather than the royal ancestors, because communication with the divine world now depended not on blood ties with the ancestors but on the moral quality of the exemplary human figure, the emperor.

Second, both the textual structure of "Wuxing zhi" and the architectural structure of the ritual complex of Wang Mang represent a supreme moral cosmology of Earth, Man, and Heaven. Whereas in "Wuxing zhi" this cosmology was reproduced in the textual structure of the three realms, Wang Mang reproduced it in material form. In the ritual complex, the circle symbolized Heaven and the square symbolized Earth. By locating the *ya*-shaped ritual hall on a circle, then a square, then a circle again, Wang Mang materialized the moral cosmology, and performed it through his ritual actions. Through the emperor's movements within, the hall of the Five Phases became an extension of the microcosm of the emperor's body, subjected to the supreme authority of Heaven, and mediating between Heaven and Earth. Wang Mang's ritual complex, therefore, submitted the hall of the Five Phases and the emperor occupying it to the authority of Heaven (symbolized by a circle) and Earth (symbolized by a square), materializing the moral cosmology represented in the textual structure of "Wuxing zhi."

Now we can conclude that the recurring cosmological pattern of *Wuxing* does not suggest a permanent cultural form, but rather an incarnation of the social conflicts that continuously transformed it. Nor is the two-thousand-year-old tradition of emperorship a unitary political institution, for it symbolizes political unity only by embodying its tensions. The two domains, cosmology as a domain of culture and empire as a domain of politics, constituted one another in a dynamic process of change.

5
Contesting Emperorship: The Center of the Cosmos and Pivot of Power

Introduction

The preceding chapter revealed the historical process of moralizing cosmology and transforming imperial sovereignty, as epitomized by the structure of "Wuxing zhi." This chapter focuses on the content of "Wuxing zhi" – that is, the various interpretations of omens. The discourse of omen interpretation best explains why and how the emperor – his person, body, and behavior – became the center of cosmology and the focus of political contestation.

Within the structure of Earth, Man, and Heaven of "Wuxing zhi," Ban Gu collected 378 cases of bad omens – catastrophes and abnormal phenomena – along with interpretations by major scholars of the Former Han. Even though the scholars represented in "Wuxing zhi" often contradict one another in their interpretations, there is common ground for all of them; all designate the emperor the center of the cosmos and the agent responsible for the bad omens caused by human disorders. All these interpreters focus their criticism on the emperor – his intentionality, behavior, and most important, his relations with various factions of power. Based on this shared focus on the emperor's person, former scholarship has read omen interpretation as a form of political criticism, checking and restraining the emperor. While this view remains highly illuminating, it carries the implication that the emperor was the possessor of power and that omens were a means of controlling that power.

This study questions the reduction of a complex cosmological discourse, that of the resonance between Heaven and Man verified through omens, to a simple dichotomy between the monarch as power holder and omens as political tools to criticize the monarch. Such a reduction reflects a view of political power as a preexisting essence that was independent of cultural and symbolic production. The alternative mode of interpretation that I propose views the cosmology of the resonance between Heaven and Man, manifested through omens, as a common discourse or constitutive symbolic structure that the Han political actors used to produce and reproduce the institution of emperorship and the concept of sovereignty, and to channel and regulate the operation of political power.

As an intrinsic aspect of political power, omen discourse inevitably embodied the vital political contest waged over the nature and function of emperorship. "Wuxing zhi," nonetheless, reflected such tension only to a certain degree, because the divergent interpretations it recorded were selected only from among those of orthodox Han Confucianism. Other, heterodox theories of omens were excluded from the text. By reconstructing some of the debates between the orthodoxy views and opposing theories excluded from "Wuxing zhi," this chapter recovers the symbolic battles over emperorship that were carried out in omen discourse. Through these reconstructions, this chapter reexamines emperorship in China's imperial history, questioning the function of the emperor in the political process and his position in the overall network of power relations during the period of imperial formation.

The central argument of this chapter is that emperorship in the Han Dynasty was in fact not the "source" of power, but rather the institution through which diverse compositions of power were tied into a body of relations. According to this argument, "political power" is no longer a self-evident entity, but rather a set of complex and dynamic relations.[1] Chinese "emperorship," the epitome of such power, did not possess all under Heaven or constitute a form of "Oriental despotism." Rather, the emperor was the pivot of power through which rival factions contested and constrained one another. Thus emperorship became a field for dispute between the military (*wu* 武) and the literati-bureaucrats (*wen* 文), between the royal lineage and the consort families, between the central government and local powers, between the imperial family and the civil government, and so on.

Omens: Magical Powers versus Heaven's Moral Intentions

Han omen discourse reflects a heightened tension between the ruler as the responsible agent for cosmic-social order and the scholar-officials as prescribers of this order. The underlying principle of omen discourse

1 Dreyfus and Rabinow, (1982) 1983, pp. 217, 221.

was that human behavior that violated the cosmic-social order would cause abnormal phenomena in both the natural and human worlds as signs of an imbalance in cosmic energies. But not all members of the hierarchical society of Han China could be connected equally with Heaven, so an exemplary human figure, the emperor – the Son of Heaven – was defined as the agent who connected Heaven and Man, and who was responsible for the total cosmic-social order. The allocation of an exemplary human figure as such a responsible agent was a locus of political dispute in the Han empire. The emperor, however, did not monopolize the knowledge and authority required to interpret omens and to define the "Way of Heaven" – the cosmic-social order. Such authority and knowledge of the Way of Heaven lay beyond the exemplary figure himself, becoming the capital of the scholars who compiled, interpreted, and transmitted texts. They occupied the civil offices of the Han Empire, operating the state bureaucracy and constructing and defining imperial institutions with their knowledge and authority.

The intense contest over emperorship between political power and moral authority was carried out among various political forces. One such contest was that between scholar-officials and religious specialists serving the emperor's person. Scholar-officials were the builders and living components of the imperial government; they were the ones who constructed the comprehensive cosmological system that sanctioned the emperor's authority yet at the same time constrained his personal power. The emperor, while dependent on the scholar-officials' system building for his authority and rule, sought to break away from the constraints of the system and increase his personal power. To do so, the emperor relied on religious specialists whom he personally hired – diviners, magic masters (*fangshi* 方士, recipe gentlemen), and shamans (*wu* 巫) – who served the emperor by helping him establish personal contact with the incalculable and infinite divine forces and deities. Benjamin Schwartz points out that the First Emperor of Qin and Emperor Wu of Han both needed yet resisted the scholars' systematization of cosmology and "subordination of divergent divine powers to the constraints of a single system." On the one hand, they needed correlative cosmological systems to ensure the support of cosmic forces for their dynasties. On the other hand, however, both emperors were eager to establish direct personal contact with the divine by using "shaman/magicians" who claimed direct access to the deities and divine powers.[2]

Scholar-officials constructed emperorship as a symbol of unity and centrality, a fundamental institution for unifying the diverse elements of

2 Schwartz, 1985, pp. 374–75.

power into a centralized state and subordinating diverse cultural and religious practices to a single dominant moral cosmology. By contrast, the religious specialists represented resistance to this unification on the part of both the emperor and regional popular cultures. This resistance took the form of seeking personal contact with the divine world and magic power in order to escape the constraints of a unified system of moral cosmology. The Han scholars' painstaking theorization and systematization of the practice of omen interpretation, and the resistance they constantly encountered, epitomize such conflict.

Interpreting catastrophes and strange phenomena was a mantic practice that had existed before the imperial era. But omen interpretation at that time was not only unsystematic, it was also used for fortune telling and prohibitions, with no clear link to the moral intention of Heaven. Strange and catastrophic natural phenomena had been seen as bad omens as early as the time of the fourth century B.C., as reflected in *Zuozhuan*. But the omens in *Zuozhuan* were still interpreted in an unsystematic manner, inconsistently related to *Yin-Yang* or to the divination of good and bad fortune. For example, in the sixteenth year of Duke Xi, five stone meteorites fell from the sky and six birds flew backward crossing the Song capital. Duke Xiang of Song asked scribe Shu Xing from the Zhou court: "What do these omens mean? Where will the auspicious and ill fortune be?" The scribe, while giving his prediction that ill fortune would fall on various states, made clear that such omens belonged to the category of *Yin* and *Yang*, the work of nature, rather than to the category of auspicious or ill fortune brought about by Man.[3] This case indicates that for Warring States rulers and religious specialists, omen interpretation was a way of predicting fortune based on natural phenomena.

By the beginning of the Han Dynasty, the interpretation of omens had become a popular mantic practice. The *Yin-Yang* texts uncovered at Yinqueshan 銀雀山 and dated to 134 B.C. contain both good and bad omens. These omens verify the cycles of the four seasons and the conquest of *Wuxing*, rather than the moral intentions of Heaven as seen in "Wuxing zhi." The auspicious omens – the phoenix, sweet spring, the wandering dragon – were no more than verifications that seasonal prohibitions, such as not cutting down trees in the spring and not offending water in the winter, were being followed. There was no connection between these omens and the moral quality of the sovereign or the Mandate of Heaven.[4] Similarly, bad omens – such as various

3 *Zuozhuan*, Xi 16; references are to Ruan Yuan, 1980 (reprint).

4 Wu Jiulong, 1985, slip nos. 0260, 0619. For the translation and analysis of these Yinqueshan *Yin-Yang* texts, see Yates, 1994b, pp. 99–100.

disasters caused by "feathered creatures," "haired insects," "worms," "shell insects," and categories similar to the ones found in "Wuxing zhi" (see Table 4.6, Chapter 4) – were seen as the result of failing to give the order appropriate to each season. The five types of orders, arranged in a conquest cycle of *Wuxing*, included both "virtuous (*de* 德)," "righteous (*yi* 義)," and "beneficent (*hui* 惠)" orders that assisted birth and growth in the spring and summer, and the orders of "punishment (*fa* 罰)" and "terror (*wei* 威)" that assisted the suppression and killing of autumn and winter.[5] Here again the omens were signs verifying a natural order of the seasons in which benevolence and violence were equally important and mutually dependent, like the work of the four seasons and the five conquering powers.

In "Wuxing zhi," however, the scholars changed omens from signs of natural order or destiny to indications of Heaven's intentions, classifying all kinds of omens into the systematic moral cosmology of Heaven. In this way, the scholars claimed the highest moral authority in omen interpretation, over that of diviners and religious specialists, and turned omens from a mantic practice into a symbolic system used in constructing emperorship and a discourse for political persuasion. This systematized moral cosmology of Heaven and the monopolized authority of interpreting Heaven's intention was a form of symbolic domination by the scholars. To construct the empire as a centralized bureaucratic state, the scholars had to unite the diverse fragments of society through a single political system and a unified cosmology. Systematizing various symbolic systems and subordinating vastly divergent deities to a moral cosmology of Heaven was the scholars' way of achieving such empire building.

Lu Jia was the first Han scholar to systematically challenge the use of omens in popular practice by diviners, shamans, and *fangshi*. While the theory of omens has usually been attributed to Dong Zhongshu, it is worth noting that it was Lu Jia who established the basic principles of omen interpretation for Han scholars. In his *Xinyu*, Lu Jia established the two basic principles of orthodox omen theory. The first principle is that omens are messages from the highest moral authority – Heaven. Lu Jia states that omens are not signs of fortune or destiny, but rather messages from Heaven: "Heaven communicates with human beings by rectifying them with catastrophes and changes (*zaibian* 災變), informing them with favorable signs (*zhenxiang* 禎祥)."[6]

Because omens embody the moral intentions of Heaven, they shall

5 Wu Jiulong, 1985, slip nos. 0732, 0819, 2003, 2475, 0235, 0362, 2042, 0618, 0230, 0755, 3173, 0730; and Yates, 1994b, pp. 136–7.
6 *Xinyu* 1, p. 1.

only be interpreted according to the classics written by the sages, rather than by religious specialists. The sages comprehended these messages by "cultivating their emotions and nature, and manifesting benevolence and righteousness,"[7] and then recorded these messages in the five classics.[8] Thus the moral authority that determines the meaning of omens lies in Heaven, the sages, and the classics. And what qualifies an interpreter of omens is not his professional knowledge but his moral cultivation. Lu Jia condemns the practice of religious specialists who interpret omens based on mantic charts or instruments rather than on the classics:[9]

> Without studying the *Book of Poetry* and the *Book of Documents* and practicing benevolence and righteousness, . . . the contemporary people draw the shape of Heaven and Earth and talk about the odd indications of catastrophes and changes, . . . tempting people with fallacious changes, and surprising people with absurdity and irregularity.

The purpose of reading omens was also redefined by Lu Jia. Instead of predicting fortune or following prohibitions in order to achieve success, omen reading aimed at prescribing and correcting the conduct of the sovereign according to ethical norms. Seeing omens, "a worthy ruler should be intelligent enough to know how to correct himself, and to test [his behavior] according to the correlated categories [of the omens]."[10] Lu's theory therefore subordinated the popular practice of using omens for fortune telling and prohibitions to the moral cosmology of Heaven.

The second principle of omens that Lu Jia established, directly derived from the first, is that the ruler was the only agent responsible for bad omens. Lu Jia argues that good or evil fortune and success or failure are determined not by the natural order of the universe but by the moral conduct of the ruler. Lu Jia thus positions the ruler at the center of both the political system and the cosmos. This double centrality of the emperor subjected him to confinement by both the bureaucratic and cosmological systems, with his behavior constantly prescribed and checked by omens:

> The decline of the world and the loss of *Dao* is not what Heaven makes happen, but rather what the ruler of the state causes to happen. Evil government generates evil *qi*, and evil *qi* induces catastrophes and abnormal phenomena. . . .

7 Ibid., p. 2.

8 Lu Jia not only advocated the study of the five classics, but also quoted extensively from *Chunqiu* and the *Guliang Commentary* on a wide range of omens and their meanings. *Xinyu* 2, p. 4; 4, p. 7; 8, p. 14; 11, pp. 18–19.

9 *Xinyu* 9, pp. 15–16.

10 Ibid., pp. 15–16.

When the *Dao* of ruling is missing below, the pattern of Heaven will reflect it above. When evil government spreads through the people, insect plagues will be generated on Earth.

In Lu Jia's theory, the ruler becomes the sole agent responsible for the resonance between Heaven and Man, and omens signal this resonance. The cosmological explanation of resonance is *qi*, which is not a natural energy but rather an embodiment of the moral value of good and evil. Evil conduct by the ruler produces evil *qi*, which in turn induces evil omens; while virtuous kings produce good *qi* that induces favorable omens. In this way, Lu Jia converts catastrophes and strange phenomena from signs of a natural order of the cosmos into signs of deviation from the moral order of Heaven. After Lu Jia established the basic principles of omen theory, scholars used omens to represent a moral authority that was above political power, prescribing norms for the performance of the ruler.

Lu Jia's theory, therefore, directly challenged the authority of religious specialists who served the emperor in the emperor's personal endeavor to contact divine forces. The more aggressive emperors of Qin and Han – the First Emperor of Qin, the founding emperor Gao, and the most flamboyant emperor of Han, Emperor Wu – all eagerly used various kinds of diviners, shamans (*wu*), and spiritual "intermediaries" (*fangshi*) to establish direct personal contact with the world of numinous spirits and mechanical forces. According to *Shiji*, the First Emperor of Qin worshipped eight deities and five *Di*, using shamans in ritual performance and using *fangshi* in attempts to contact immortals and achieve immortality. Emperor Gao collected shamans from various regions representing different religious traditions, using them to perform sacrificial rituals to a much greater number of deities. Emperor Wu was particularly preoccupied with his personal contact with the unlimited and divine powers. His court attracted numerous shamans and *fangshi*, some of whom were given tremendous power, wealth, and prestige. Emperor Wu also had many temples and altars built for the worship of various spirits and divine powers.[11] It was this practice of the emperors that empowered religious specialists such as shamans and *fangshi*, and that directly conflicted with the scholars' moral authority.

To constrain the emperor's person, to construct emperorship as the core institution of empire building, and to monopolize divine and moral authority, the scholars used omen theory as a major symbolic resource. Lu Jia's theory established basic principles for omen interpretation – a

11 *SJHK* 28, pp. 21–3, 36–8, 45–88; *SJ*, pp. 1367–8, 1378–9, 1384–404.

single, unified moral cosmology of Heaven, with omens serving as verifications of Heaven's moral intentions. Yet Lu Jia died without seeing his theory become a reality. It was Dong Zhongshu who made these principles orthodox and who applied the theory to the political practice of interpreting omens in memorials submitted directly to the emperor.

The Cosmological Debate over Emperorship

Besides the resistance to systematization and centralization encountered from the emperor's person and from religious specialists of popular local cultures, the scholar-bureaucrats faced other rivals for authority and power. The most prominent rivals during the first century of the empire were members of the royal family who became strong local kings and resisted the political control of the central bureaucratic state and the ideological domination of orthodox Confucianism.

Scholars versus the Noble Kings

Noble kings played a crucial role in the stability of the early Han. The greatest threat to the Han Empire at its inception was posed by the founding emperor's confederates, who had possessed large territories and claimed themselves kings during the civil war. Emperor Gao, still in need of their support, confirmed the existing kings. By 202 B.C., ten such kingdoms had been established, controlling a larger area and even a larger population than the commanderies of the central government. It was with the royal nobility – the emperor's own brothers and sons – that Emperor Gao soon replaced the confederate kings, hoping that family ties would make these noble kings loyal. Consequently, by 196 B.C., all except one of the confederate kings had been replaced by brothers or sons of the emperor, who ruled two-thirds of the empire (Map 1).[12] For the founding emperor, no confederates or bureaucrats were as dependable as his own flesh and blood.

These noble kings, however, were the primary rivals of the scholar-statesmen, for the scholars' vision of the empire was a centralized bureaucratic state administered by scholars qualified for civil service through learning and moral quality rather than inherited status. The scholars' participation in government was institutionalized as early as the Qin Dynasty as part of the bureaucratic system that appointed "academicians" (*boshi* 博士), who were paid members of the bureaucratic gov-

12 Loewe, 1986b, pp. 124, 126.

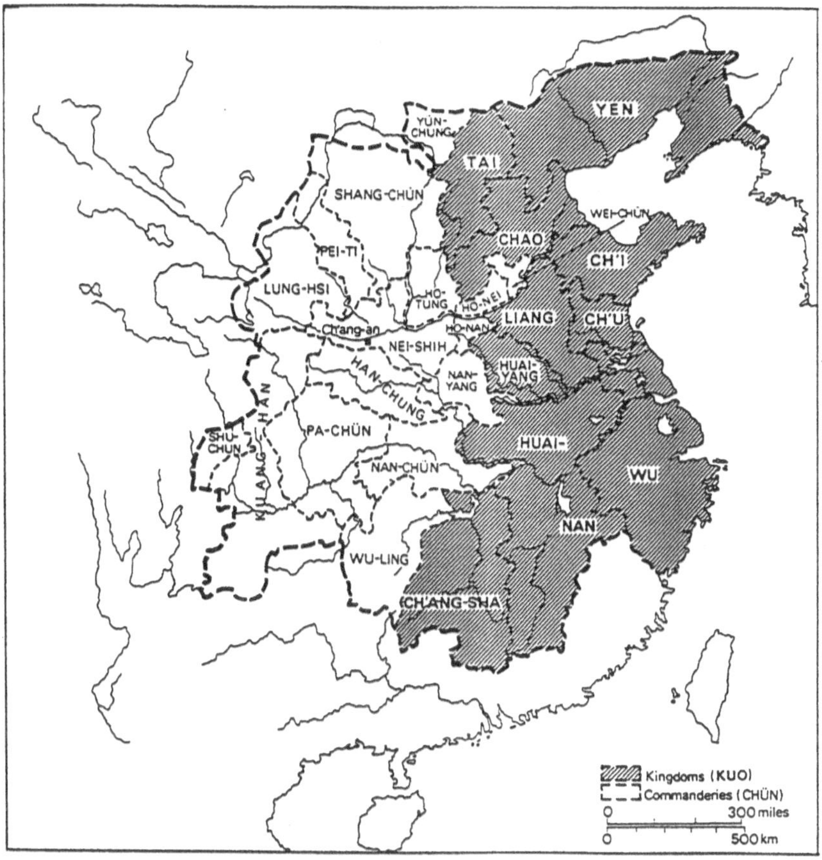

Map 1. The Han Empire, 195 Map B.C. (Twitchett and Loewe, eds., *The Cambridge History of China*, vol. 1, Cambridge University Press, 1986, p. 125).

ernment.[13] During the Han Dynasty, scholars became a political force that increasingly monopolized the bureaucratic system. This change was marked first by the establishment of an imperial academy (*taixue* 太學) under the reign of Emperor Wu (141–87 B.C.), which increased the

13 The institution of academicians was established as early as the late Warring States period, as exemplified by the Jixia 稷下 academy of Qi. Yet the imperial academy of Qin differed from the Jixia academy. According to Yu Yingshi, the Jixia academy was characterized by "critiquing yet not participating in the government," and the position of academicians was not yet part of the bureaucratic system. Starting with the Qin Dynasty, the academy became a part of the bureaucratic system of the state, and the academicians lost their independence from the state and became paid officials of the government. See Yu Yingshi, 1987, pp. 51–68.

number of academicians. Furthermore, beginning in 124 B.C., students were placed under the charge of each academician, receiving payment from the government and serving as a pool for the recruitment of civil officials. The number of students increased from fifty for each academician at the beginning, to three thousand during the reign of Emperor Cheng (23–7 B.C.).

Along with the increasing importance of the imperial academy, the recruitment of civil servants mainly depended on the recommendations of provincial officials or of senior ministers of the central government. The criterion for recommendation, as defined in numerous imperial edicts, was primarily the quality of the man's moral and literary education.[14] According to *Hanshu*, Emperor Wu adopted the advice of Gongsun Hong 公孫弘 to appoint scholars of Confucian texts to important civil service positions, and to replace clerks with such scholars. After that, the offices of the government, ranging from the highest (*gong* 公 and *qing* 卿) to the lowest (*shi* 士 and *li* 吏) were occupied mostly by "refined and sophisticated literary scholars."[15] The advocate of this policy, Gong Sunhong, himself was promoted from literary scholar (*wenxue* 文學) to academician, and finally to the highest office of the central government, chancellor (*chengxiang* 丞相).[16]

The scholars, therefore, became living components of the centralized empire, and their political life depended on the bureaucratic system. They were the "natural enemy" of the royal nobility, whose authority was based on birthright and whose power depended on political pluralism and localism. The competition between centralization and pluralism, between the scholars controlling the bureaucratic state and the royal nobility controlling two-thirds of the territory and population, was decisive for the nature and future of the empire.

For almost a century, the scholars and nobles competed with each other through their influence on the emperor and their definitions of emperorship. Since both parties had to persuade the emperor in order to eliminate their rivals and implement their own blueprints for the empire, the logic of persuasion became their most crucial discursive power. The logic of persuasion used by the scholars was traditionally referred to as *ru* 儒 or Confucianism, and that employed by the local kings was often identified as Dao 道 or Huang-Lao 黃老.

However, I choose not to use those labels for the present study. Such

14 For detailed studies of the imperial academy and the scholars' role in government, see Bielenstein, 1980; Qian Mu, 1958, pp. 165–233; Lü Simian, (1947) 1986, pp. 646–63; and Loewe, 1986d, pp. 463–6.

15 *HSBZ* 88, pp. 3b–6a; *HS*, pp. 3593–6.

16 *HSBZ* 58, pp. 1a–8a; *HS*, pp. 2613–23.

labels were relevant only to the degree that the political actors claimed them as sources of authority, with one side appealing to the authority of the Confucian sages and the other appealing to the tradition of Daoist sages. The conceptual boundaries of these theoretical traditions in Han times were in fact extremely difficult to define, and the political boundaries of the people who mobilized these traditions by no means coincided with the conceptual boundaries; the same people could mobilize different traditions, and the same tradition was often used by different people representing different political positions. Since this study is concerned mainly with political controversies rather than with reconstructing "schools" of thought, such terms might be more confusing than helpful for this particular analysis.[17]

Dong Zhongshu and the King of Huainan

The diametric opposition between Dong Zhongshu (179–104 B.C.) and the second king of Huainan, Liu An (d. 122 B.C.), best illuminates the competition between these two political forces. Dong's arguments were mostly made to Emperor Wu in his memorials, which are recorded in *Hanshu.* While scholars in the capital, such as Dong Zhongshu, influenced the emperor by submitting memorials or being called in for consultations, the kings of distant kingdoms did not have such channels of influence. The king of Huainan spoke to the same emperor by submitting to him a book that the king and his circle of consultants had written collectively, which was later titled *Huainanzi.*

Dong Zhongshu's political position has been a point of scholarly dispute. Some scholars, such as Feng Youlan, see Dong Zhongshu's cosmological and political theories as ideological justification for authoritarianism and the despotic power of the emperor.[18] Other scholars, exemplified by Hsiao Kung-chuan, point out that Dong's theory was after all aimed at using the authority of Heaven to limit and constrain the emperor's despotic desires.[19] Sarah Queen synthesizes both views of Dong – as a faithful ideologue for the emperor and as a critic of the throne – to depict Dong's thought as being "complex and contradictory."[20] Gary Arbuckle, by contrast, explains such contradiction by depicting Dong as an idealist who looked beyond the Han to a new sage-king to take over the Han throne.[21] I am convinced that Dong's

17 For the problems of the terminology of Confucianism, Daoism, Legalism, and Huang-Lao, see Loewe, 1995.
18 Feng Youlan, 1983–4, vol. 3, pp. 39–89.
19 Hsiao Kung-chuan, vol. 1, trans. F. W. Mote, 1979, p. 487.
20 Queen, 1996, p. 4. 21 Arbuckle, 1995.

Confucian doctrines connot be interpreted as a black-and-white matter of supporting versus restraining the despotic power of the emperor. Dong's agenda was to construct imperial sovereignty, to define the norms of political performance of the emperor, and ultimately, through the construction of emperorship and persuasion of the emperor, to eliminate rival political forces and prescribe an entire sociopolitical order for the empire.

Huainanzi, a great scholarly achievement that has been praised as an encyclopedia of knowledge of the time, was also written as a blueprint for the empire and emperorship.[22] This is stated explicitly in the concluding chapter of the book:[23]

> The book of the Liu family (*Huainanzi*) observes the image of Heaven and Earth, connects the affairs of the ancient and present times, weighs these affairs in order to establish institutions and rules, and models [the images and affairs] in order to apply [the rules] appropriately.

By calling it "the book of the Liu family," Liu An had in mind his most important royal relative, the emperor, as the reader; a profound exposition of cosmology and history was presented as a model for the emperor to use in his governing.

Huainanzi has been increasingly recognized as a Daoist, or Huang-Lao Daoist work. The two most recent studies of *Huainanzi*, by Harold D. Roth and John S. Major, support the view that *Huainanzi* "is a principle representative of Huang-Lao thought during the Han."[24] Major further explains that the differences between the theory of *Huainanzi* and that of Dong Zhongshu's Confucian doctrines reflect the fact that *Huainanzi* predated the formulation of the Han Confucian synthesis led by Dong Zhongshu.[25]

While many scholars have accepted that *Huainanzi* represents the Huang-Lao school of thought that dominated imperial ideology before the establishment of Han Confucian doctrines, there are other viewpoints. Roger Ames sees Chapter 9 of *Huainanzi* as representing not a dominant court ideology but rather an alternative to the totalitarian control of the imperial court.[26] Michael Loewe further suggests that it

22 *Huainanzi* covers almost all branches of knowledge existing at the time, ranging from cosmology, astronomy, geography, and epistemology to human nature, rulership, government, military strategy, and human society and customs. For textual analysis of the book, its authorship and transmission, see Roth, 1992.

23 *Huainanzi* references are to Liu Wendian, *Huainan honglie jijie* (abbreviation *HNHLJJ*) 21, pp. 9b–10a.

24 Roth, p. 13; and Major, 1993, pp. 8–14.

25 Major, 1993, pp. 5, 14.

26 Ames, 1983, p. xvi.

was possible that certain parts of *Huainanzi*, written after 139 B.C., were meant to voice an uncomfortable word of protest rather than ideological support for the policies of the central government.[27]

The exact dating of *Huainanzi* is a problem I will not try to solve. The relation of *Huainanzi* to the "Huang-Lao" school of thought is also beyond the scope of the present study, since it is still open to question whether or not Huang-Lao was a single ideology and what its tenets were. In the remaining part of the chapter, I will concentrate on the political and theoretical conflicts between the king of Huainan and Dong Zhongshu, seeing the two men as representatives of two contemporary rival forces rather than of two orthodox ideologies, one preceding the other. The theoretical opposition between Dong Zhongshu and the king of Huainan epitomized the symbolic and political struggles between the centralized empire represented by scholar-officials and the pluralism represented by noble kings. While the controversies between Dong and Liu An covered a vast range of topics, this chapter focuses only on their symbolic battle over the cosmological construction of two conflicting models of emperorship.

The Symbolic Battle over Cosmology

The debate over two models of emperorship, represented by these two men, was carried out in the cosmological discourse, and the key to this debate is the concept of resonance between Heaven and Man (*tianren ganying* 天人感應). While both men put the concept of resonance at the center of their cosmology, their definitions of the term differ sharply.

Dong Zhongshu fully developed Lu Jia's moral cosmology and omen theory, elaborating on the idea of Heaven as an anthropomorphic deity and omens as Heaven's speech to the ruler. This theory is stated in the opening of the first of Dong Zhongshu's three memorials responding to Emperor Wu's inquiry:[28]

> The subject of your Majesty [i.e., Dong himself] prudently investigated [the cases in] the *Spring and Autumn* (*Chunqiu*), observing the affairs of the former dynasties; accordingly I see the Heaven and Man connection, which is extremely awe-inspiring. Whenever a state will decline due to the loss of *Dao*, Heaven will first give catastrophes to warn [the ruler]. If [the ruler] fails to realize and introspect, [Heaven] will further show strange phenomena to scare the ruler. If the ruler still does not change [his conduct], then damage and decline will follow. From this one sees that the heart of Heaven kindly loves the human ruler; he intends to stop the chaos for the ruler.

27 Loewe, 1995.
28 *HSBZ* 56, p. 3a; *HS*, p. 2498.

Dong here strongly opposes all interpretations of cosmos as a natural or physical order free from moral intention. From Zhou theology he revives the idea of Heaven as an anthropomorphic deity, attributing to it a heart, intention, and love. Heaven manifests its will in omens, warning the ruler of the human world. Dong thus defines omens as Heaven's speech. And it is Heaven, the highest moral consciousness, that makes catastrophes and strange phenomena happen.

The king of Huainan and his circle advocate the very opposite of Dong's theory. They see the same phenomena as the result of resonance (*ganying* 感應), that is, phenomena of the same category and sharing the same *qi* affecting one another mechanically:[29]

When the *Yang qi* prevails, it scatters to make dew;
When the *Yin qi* prevails, it freezes to make frost and snow.
Creatures with feathers make up the category of flying things, and are subject to *Yang*.
Creatures with shells and scales make up the category of creeping and hiding things, and are subject to *Yin*.
. . . .
Fire flies upward, water flows downward; thus
The flight of birds is aloft, the movement of fishes is downward.
Things within the same category mutually move each other.

Here *Huainanzi* replaces Dong's cosmology of the moral intention of Heaven with a physical cosmology of resonance – "resonance within the same category" or "resonance with the same *qi*."[30]

The two opposing theories of "resonance" – physical resonance in nature versus Heaven's speech – were directly linked to the conflict over the authority to interpret omens: where such authority lay and who possessed it. In Dong's theory, since omens are Heaven's speech, the translation of this divine speech into human knowledge is the key to authority in the human world. Dong argues that the only human being who could comprehend the divine messages is the sage Confucius, who converted the meaning of the omens into a sacred text, the *Spring and Autumn*:[31]

The subtle [communication] between Heaven and Man is the *Dao* of the ancient and the present. Confucius compiled the *Spring and Autumn*, examining the *Dao* of Heaven above and inquiring into the human sentiment below, referring to the ancient and investigating the present. That is why what the *Spring and Autumn* criticizes is what catastrophes are imposed upon, what the *Spring and Autumn* dislikes is what strange phenomena are given for. It writes about the mis-

29 *HNHLJJ* 3, pp. 2a–3a; translation adopted with modification from Major, 1993, p. 65.
30 *HNHLJJ* 6, p. 4b. For the study of the concept of resonance in *Huainanzi*, see Le Blanc, 1985.
31 *HSBZ* 56, p. 14b; *HS*, p. 2515.

takes of states and families, and the changes of catastrophes and strange phenomena, in order to show that what Man does and the extremities of good and evil resonate back and forth and communicate with Heaven and Earth.

Intervening between the moral intention of Heaven and the conduct of the ruler, Dong here introduces the concept of the sage and the sacred text. He implies that the authority for translating Heaven's speech should lie not in any other profession, knowledge, or tradition, but only in the sage Confucius and his writing. Accordingly, the scholars who have mastered the sage's writings are the only living men with the authority to interpret omens.

Huainanzi, with its concept of "resonance with the same *qi,*" directly challenges this monopolization of sacred knowledge and authority. It rejects the use of any knowledge or moral values in interpreting the resonance between things. It argues that the resonance of things cannot be subject to moral judgment or knowledge: "The resonance of things belonging to the same category is darkly mysterious and extremely subtle. Knowledge cannot explain it, nor discussion unravel it."[32] Therefore, omens as the signs of resonance are inexplicable and mysterious, free of any explicit moral meaning.

The symbolic battle between the two theories of resonance was extended to the entire cosmology, involving most of the cosmological concepts widely used in both philosophical writings and popular culture of the time.[33] Crucial concepts brought into the symbolic battle included *Dao,* Heaven, *qi,* and *Yin-Yang.*

Although promoting his own theology of Heaven and of omens as Heaven's speech, Dong did not, or could not, disregard these widely used concepts. He instead integrated them into his cosmology of Heaven, and subordinated them to the moral evaluation of Heaven. The two conflicting theories, therefore, came to share the same vocabulary of *Dao,* Heaven, *Yin-Yang, qi,* and resonance, while fundamentally opposing one another in defining these terms and the cosmology they represented.

In the same memorial, Dong also treated omens as resonance through the same *qi* and within *Yin* and *Yang* categories. But the concepts of *qi* and *Yin-Yang* in Dong's theory again served as manifestations of the moral value of Heaven:[34]

32 *HNHLJJ* 6, p. 3a; translation adopted with modification from Le Blanc, 1985, p. 116.

33 The popularity of these concepts can been seen in the archaeologically discovered texts dating from the late Warring States period to the early Han Dynasty. Representatives of such texts are the almanacs, divination manuals, military texts, and medical texts discussed in Chapter 3, and the funeral library of Mawangdui tomb number three.

34 *HSBZ* 56, p. 4b; *HS,* p. 2500.

[The good omens that appeared in ancient times] were all results of accumulating good and gathering virtue. Up to the later reigns, [the rulers] were lewd, unrestrained, declining, and weak. They could not govern the population, then the princes revolted, and the bandits and good people fought for land. They abandoned moral education and used punishment. When punishment was applied unjustly, evil *qi* accumulated below, and complaints and hatred gathered above. The above and the below were not in harmony; that is why *Yin* and *Yang* were in conflict and goblins and plagues were thus produced. These are the causes of the catastrophes and strange phenomena.

Here Dong distinguishes *qi* according to the moral value of good and evil, and judges resonance between conduct and omens also according to moral standards. Thus the inexplicable "resonance of things within the same category" in *Huainanzi* is replaced by the explicable moral resonance of good and evil.

Besides the concepts of resonance and *qi*, the two theories also oppose one another regarding the meaning of, and the relation between, *Dao* and Heaven. For the king of Huainan and his followers, *Dao* is superior to Heaven. The superiority of *Dao* over Heaven is stated in the cosmogony in *Huainanzi*:[35]

The *Dao* began in the Nebulous Void.
The Nebulous Void produced spacetime;
Spacetime produced the primordial *qi*;
A shoreline (divided) the primordial *qi*.
That which was pure and bright spread out to form Heaven;
The heavy and turbid congealed to form Earth.
. . . .
The conjoined essences of Heaven and Earth produced *Yin* and *Yang*.
The supercessive essences of *Yin* and *Yang* caused the four seasons.
The scattered essences of the four seasons created all things.

In *Huainanzi's* cosmogony, *Dao* not only is given ontological priority, but also is portrayed as the dynamism, the operating force and governing order of the cosmos. Thus *Dao* is the highest authority above all things: "*Dao* covers Heaven and supports Earth . . . with its power embraces Heaven and Earth, and harmonizes *Yin* and *Yang*, regulating the four seasons, and ordering the Five Phases."[36] Heaven here is an aspect of nature rather than a deity; it is the celestial part of the cosmos complemented by its counterpart, Earth. It is subordinate to *Dao*, being generated and operated by *Dao*.

Dong Zhongshu, by contrast, not only gives Heaven human inten-

35 *HNHLJJ* 3, pp. 1a–b; translation adopted from Major, 1993, p. 62.
36 *HNHLJJ* 1, pp. 1a–2a.

tions, emotions, and purposes, turning it from a celestial part of nature into a deity, but also reverses the hierarchy of *Dao* and Heaven: "The greatness of *Dao* lies in its origin in Heaven."[37] He gives Heaven the highest ontological priority and designates Heaven the active agent that makes things happen in the universe. Besides reversing the hierarchy between *Dao* and Heaven, Dong also redefines the properties of *Dao*. While in *Huainanzi*, *Dao* is the cosmic order that transcends human values, Dong redefines *Dao* as the sociopolitical order of human beings, with the moral principles of Confucius's teaching as its content: "*Dao* is the Way that is appropriate for government, with benevolence (*ren* 仁), righteousness (*yi* 義), rites (*li* 禮), and music (*yue* 樂) as its instruments."[38]

A similar symbolic battle was being waged over the meaning of *Yin* and *Yang*. In the cosmogony from *Huainanzi* quoted earlier, *Yin* and *Yang* are the stuff and energy that compose the physical universe, and the distinction between the two comes from their physical natures – "pure and bright" (*qing* 清) versus "heavy and turbid" (*zhuo* 濁). *Yin* and *Yang* interact, thus producing movement and transformation in the cosmos, and there is neither domination nor moral superiority of one over the other. Therefore, *Yin-Yang* movement represents the dynamic and nonhierarchical transformation and resonance of ten thousand things in the universe, all of which originated from the same *Dao*.

By contrast, Dong employs the concept of *Yin* and *Yang* to embody the moral value of Heaven rather than the physical nature of the cosmos, with *Yang*, representing the moral value of Heaven, considered far superior to *Yin*, which represents punishment. Dong thus orders *Yin* and *Yang* in a fixed hierarchy, in which the superior *Yang* dominates the inferior *Yin*:[39]

> The greatness of the *Dao* of Heaven is in *Yin* and *Yang*. *Yang* is bounty (*de* 德), *Yin* is punishment (*xing* 刑). Punishment functions to kill; bounties function to give birth. That is why *Yang* permanently resides over the peak of summer and is in charge of birth, nurturing, cultivation, and growth, and *Yin* permanently resides over the peak of winter and accumulates at the place of emptiness, void, and uselessness. This makes clear Heaven's appointing virtue and dismissing punishment. Heaven orders *Yang* to come out to distribute his offers from above and to be in charge of accomplishing the year, and orders *Yin* to stay in and subside below, sometimes coming out to assist *Yang*. . . . This is the will of Heaven. The ruler receives the will of Heaven and arranges business accordingly. That is why [he shall] use moral education, and not use punishment.

37 *HSBZ* 56, p. 16b; *HS*, p. 2518–9.
38 *HSBZ* 56, p. 18b; *HS*, p. 2499.
39 *HSBZ* 56, pp. 5a–b; *HS*, p. 2502.

The symbolic dispute between the two conflicting cosmologies, as reconstructed so far, was carried out using a great deal of shared terminology – resonance, *Dao*, Heaven, *qi*, and *Yin-Yang*. Yet using these common terms, Dong Zhongshu and the king of Huainan each constructed a cosmology antithetical to the other's, differing over fundamental questions about the very nature of the cosmos – a supreme moral judge versus an immanent and ineffable natural order – and its relation to Man.

The Two Opposing Models of Emperorship

The symbolic battle over "resonance" in particular and cosmology in general involved the construction of two opposing models of emperorship, two views of where the emperor was located in the cosmos and what his function was in the political system of the empire.

The emperor's cosmological position was determinative of his political function. Dong Zhongshu built a political theory of sovereignty using the concept of the resonance between Heaven and Man. He defined the sovereign not as the holder of political power, but as the moral agent responsible for conveying the moral intention of Heaven into the sociopolitical order of the human world. Dong interprets the very graphic of the character for king – *wang* 王 – by saying that the three horizontal strokes symbolize the three realms of Heaven, Man, and Earth, with the central vertical stroke connecting the three to symbolize the sovereign. He defines the sovereign as the sole human being capable of achieving resonance between Heaven and Man, saying: "To be at the center of Heaven, Earth, and Man, and thus to connect the three and put them in communication, is the responsibility of the sovereign, and nobody else can bear such responsibility."[40] Therefore, "Heaven and Man resonance" (*tianren ganying*) is the function exclusively of the sovereign.

Huainanzi resists such cosmological centrality of the emperor using a pluralist model of resonance. It states that resonance has nothing to do with the status or authority of the sovereign, but can be achieved by anyone who unites with *Dao*. In Chapter 6 of *Huainanzi*, the authors present seven people who have achieved resonance with Heaven, Earth, Man, and animals; they are a blind musician, a commoner's daughter, King Wu of Zhou, a high official of Chu, a Chu music master, a renowned archer, and a famous angler. These people have achieved resonance with the universe not through their high status, nor through moral knowledge

40 *CQFLYZ* 11, "Wangdao tongsan 44," p. 9a.

or purposeful effort, but through perfect sincerity (*jingcheng* 精誠). *Huainanzi* specifically states that such resonance has nothing to do with social hierarchy or authority:[41]

> Take this blind music master and commoner's daughter; their rank was lower than that of the Director of Hemp, their authority lighter than floating feathers. Yet by concentrating their essences (*jing* 精) and disciplining their thoughts, discarding all concerns and gathering together their spirits (*shen* 神), they merged above with the Nine Heavens and stimulated their most subtle essences (*zhijing* 至精).

Such a cosmological debate over the emperor's position in the resonance between Heaven and Man was the basis for the political debate over two models of emperorship – the active sovereign of Dong Zhongshu versus the nonactive sovereign of *Huainanzi.*

The confrontation between these two models of emperorship started in the early Han. Even at a time when the Huang Lao concept of "nonaction" (*wuwei* 無為) was predominant in the court during the first few reigns, Jia Yi alerted the emperor to the lack of authority of the throne and the lack of hierarchy in society, attributing the situation to the "nonaction" model of emperorship. He cried out in his memorial to Emperor Wen: "Now society is disrespectful [to the emperor] in the extreme, lacks hierarchy in the extreme, and is offensive to the emperor in the extreme, but the advisors [of the emperor] still talk about 'nonaction'; this is what makes one sigh in despair."[42] Jia Yi died in despair, facing a situation in which the strong noble kings dominated the larger portion of the empire and the "nonaction" model of emperorship dominated the court. But with the support of Emperor Wu, who personally disliked the nonaction model, Dong Zhongshu developed Jia Yi's themes of elevating the authority of the emperor, defining social hierarchy, and establishing Confucianism as the state orthodoxy. He achieved all three by recentering emperorship in the moral cosmology of Heaven.

Positioning the emperor at the center of the cosmos as the only human agent connecting Heaven and Man, Dong Zhongshu argues that the emperor should be the active (*youwei* 有為) moral agent. The active function of the ruler derives from the active nature of Heaven. Emperorship is about the active realization of Heaven's will in the human world; the emperor must be able to "receive the action of Heaven from above, and to correct his own actions below."[43] Being the sole connec-

41 *HNHLJJ* 6, pp. 1a–b; translation adopted from Le Blanc, 1985, pp. 103–4.
42 *HSBZ* 48, p. 18b; *HS*, p. 2243.
43 *HSBZ* 56, p. 5a; *HS*, p. 2502.

tion of Heaven and Man, the emperor is the single human agent who makes things happen: "The human sovereign stands in the position of deciding to let live or to kill, sharing with Heaven the control of power over change and transformation."[44]

In contrast to Dong's active sovereign, *Huainanzi* promotes a "nonaction" model of rulership by decentralizing the emperor's position in cosmic resonance. As if written word for word in response to Dong's active model of emperorship, the authors of *Huainanzi* declare:[45]

> Does the ruler of all under Heaven have to possess the power, control the situation, hold the power of deciding to let live or to kill, so that his order shall be followed?! What I [we] mean by the sovereign of all under Heaven is not this kind, but rather the kind that fulfills one's own nature . . . and to do so means to unite with *Dao.*

Emperorship, according to *Huainanzi,* is not taking action to dominate and control, but rather uniting with *Dao* through nonaction.

The two models of emperorship oppose one another on the level of the fundamental cosmological principles of Heaven and *Dao.* Dong derives his active and willful sovereign from the active creation and intentionality of Heaven. In Dong's theory, the will of the ruler, like the will of Heaven, is the creative force that makes things happen. The ruler should rule actively with his will, because Heaven actively creates and transforms with its intentionality:[46]

> The human sovereign changes social customs and practices with his favor, dislike, happiness, and anger, while Heaven transforms grass and trees with warmth, coolness, coldness, and heat. Therefore Heaven, Earth, and the human sovereign are one.

Dong elevates the emperor to unity with Heaven, with the emperor using his intentions and emotions to control and change the human world, just as Heaven uses its intentions in the natural world.

Huainanzi sharply opposes such authority of the emperor's intentionality, stating that ideal rulership is precisely that in which the sovereign eliminates his personal intentions and emotions. It warns the ruler: "Do not take actions according to ugliness or beauty, or out of favor or dislike. Do not take actions of reward and punishment based on delight or anger. . . . Do not make things happen out of one's own intentions." What the ruler should do is "empty his heart and weaken his will."[47]

In contrast to the active Heaven on which Dong Zhongshu's active

44 *CQFLYZ* 11, "Wangdao tongsan 44," p. 11b.
45 *HNHLJJ* 1, p. 22b.
46 *CQFLYZ* 11, "Wangdao tongsan 44," p. 12a.
47 *HNHLJJ* 9, pp. 1b, 10a.

model of emperorship was founded, *Huainanzi* constructed the "nonaction" model of emperorship based on the "nonactive" *Dao.* In this "nonaction" model, Heaven and Man connect through spontaneous resonance with *Dao.* In contrast to the active creator and the purposeful Heaven, *Dao* is nonactive, nonpurposeful, and nonpossessive:[48]

> The mighty and supreme *Dao* gives birth to ten thousand things but does not possess them, accomplishes and transforms phenomena but does not dominate them. . . . That is why all affairs under Heaven cannot be made to happen, and one can only follow the natural development of them.

While many scholars have labeled *Huainanzi* a Huang Lao text, based on the terminology of *Dao* and nonaction, a major distinction between *Huainanzi* and Huang Lao thought must be made. In contrast to the Huang Lao theory of the early Han statesmen, who used the term "nonaction" to promote laissez-faire politics, the authors of *Huainanzi* constructed a comprehensive cosmological and political theory of the emperorship of "nonaction," thus systematically opposing the theory of the authoritative and active ruler. The ruler should be "nonactive" because *Dao* is nonaction.

The two models of rulership also contradict one another at the level of human nature. What Dong means by "action" (*wei* 為) is to actively realize Heaven's intention in order to fulfill human nature through morality. In his memorial, Dong points out three fundamental actions of the sovereign: to receive Heaven's moral intentions, to rule by moral education, and to establish norms and social hierarchy, all of which are essential for fulfilling human nature:[49]

> Heaven's ordinance is called Mandate, and Mandate cannot be put into operation except by men of sagely qualities. The natural quality of Man is called human nature (*xing* 性), and this cannot be fulfilled save by moral education and transformation. Human desires are called emotions (*qing* 情), and these cannot be regulated except by institutions and rules. This is why the human sovereign carefully receives the intention of Heaven from above, so that he may conform with the Mandate. Down below he strives to educate and transform his people by enlightening them, in order to fulfill their nature. He establishes norms for law and institutions, distinguishes between the upper and the lower orders of society, in order to preclude desire. If he achieves these three aims, the fundamental basis will be established.

It is precisely these three fundamental actions that the nonaction theory opposes. *Huainanzi* condemns Dong's three actions as corrupting human nature and deviating from *Dao. Huainanzi* also criticizes the

48 *HNHLJJ* 1, pp. 2b, 6a.
49 *HSBZ* 56, p. 15a; *HS*, pp. 2515–16.

fundamental Confucian principles underlying these three actions – the principles of "benevolence" and "righteousness":[50]

Today, accumulating benefits, enriching generosity, gathering affection, inheriting largess, using speech to persuade people, deceiving the people with knowledge, and making them happily indulge in their nature – all these constitute "benevolence."

To set up great fame, to realize ruler-subject distinction, to establish a hierarchy of high and low, to distinguish kin from strangers, to rank the noble and the humble, to save falling states, to preserve a decaying house, to quell turmoil, to control chaos, to restore decadent clans, to regenerate an extinct lineage – all these constitute "righteousness."

......

When benevolence and righteousness are established, then *Dao* and *de* are abandoned and lost.

Unlike Dong's theory of human nature based on Heaven's intention, *Huainanzi* proposes a cosmology of human nature in support of the idea of nonactive rulership. *Huainanzi* sees any action or external influence as harmful to inborn human nature: "Man is born quiescent, because that is the nature of Heaven. But once he is moved [in response to outward influences] to take actions, it damages the inborn nature."[51] Based on the nonactive *Dao* and the inborn human nature of quiescence, the human ruler should rule without taking actions, because only through nonaction can he be united with *Dao* and preserve the quiescent human nature: "The art of rulership is to deal with things through nonaction and to disseminate wordless instructions. Limpid and quiescent, he does not move; even when moved he is not agitated."[52]

Therefore, the two conflicting cosmologies constructed two opposing models of emperorship. Based on the moral cosmology of Heaven and resonance as Heaven's speech, modeled upon Heaven's active role as the highest moral intentionality, Dong Zhongshu's model of emperorship constructed the emperor as the sole human agent who connects Heaven and Man, who initiates and changes things in the human world according to his own moral judgments and emotional inclinations, and who fulfills the human nature that was Heaven-given by imposing on the populace institutions of education, hierarchy, and regulations.

As a contradiction and resistance to such a model of the cosmological and political centrality of emperorship, *Huainanzi*'s cosmology is that of an immanent and ineffable *Dao* that transcends all human moral values and knowledge. The ten thousand things produced by *Dao*,

50 *HNHLJJ* 2, pp. 10b–11a.
51 *HNHLJJ* 1, p. 6b.
52 *HNHLJJ* 9, p. 1a; translation is adapted from Ames, 1983, p. 168.

including human beings, embody the tranquility and spontaneity of *Dao* as their inborn nature. Since things sharing the same *qi* spontaneously move one another, there is no need for a human agent to interfere with the work of *Dao* according to moral intentions or hierarchy. The role of the human sovereign, therefore, is to imitate the nonaction of the *Dao*, to eliminate intentionality and moral judgment, to get rid of hierarchy and distinctions, and to serve as a model for fulfilling quiescent human nature.

The Political Struggle over Empire Building

The cosmological battle over emperorship was at the center of empire building in Early Han. It articulated the political struggle between the two political forces, the scholar-officials and the noble kings, over the order of the empire – centralization (*dayitong* 大一統) versus pluralism. The political struggle between centralization and pluralism was carried out in the realm of culture and ideas, as well as in the realm of political structure.

Centralization in the Realm of Culture and Ideas

In Dong Zhongshu's theory, the active ruler is essential for the cultural and ideological unification of the empire. Although Dong Zhongshu designates the emperor as the highest moral agent among human beings, he also subordinates the emperor to the moral authority of Heaven, whose speech can only be comprehended by the masters of Confucian texts. In this way, Dong claims that the text of *Spring and Autumn* is the supreme authority for ideological unification, and he proposes that Emperor Wu take action to prohibit all other schools of thought:[53]

> *The Spring and Autumn* is the supreme unifier, because it is the permanent principle of Heaven and Earth, the common ethics of the ancient and the present. But today, the masters of learning are different in their ways, the interpretations of meanings vary, that is why the emperor has nothing to hold as an orthodox unifier. . . . Your subject [i.e., Dong] suggests that all that is not in the "six arts" and the methods of Confucius should be abolished, prohibiting them from being promoted along [with Confucianism]. Only when depraved theories are exterminated and have disappeared can the universal norms be unified, the rules and standards become clear, and people know what to follow.

53 *HSBZ* 56, p. 19a; *HS*, p. 2523.

Accepting Dong's suggestion, Emperor Wu took the famous step of dismissing the hundred schools and establishing Confucianism as the state orthodoxy. In 136 B.C., Emperor Wu changed the official system of appointing academicians, establishing chairs only for the five Confucian classics.[54] Thus, by synthesizing cosmologies using the single authority of Heaven, centering the cosmos on the emperor's person, and singling out Confucianism as the only truth about Heaven, Dong Zhongshu provided a blueprint for eliminating political and ideological pluralism and truly accomplishing the ideological unification of the empire.

Huainanzi, on the eve of the prohibition of the hundred schools, protested that the hundred schools all had the right to engage in government: "The hundred rivers have different sources, but they all flow to the ocean. The hundred schools have different achievements, but they are all devoted to good government."[55] Actually, within this pluralistic view of the hundred schools, the authors of *Huainanzi* devalued the teaching of Confucian classics in particular: "It was only when the *Dao* was damaged that the *Book of Poetry* was compiled. It was only when the house of Zhou decayed and the rites and righteousness collapsed that the *Spring and Autumn* was written. Therefore what the learning of the *Book of Poetry* and *Spring and Autumn* praises are the inventions of a decaying era."[56]

The centralization of ideas and cultures and the establishment of a single imperial orthodoxy, which Dong proposed and Emperor Wu actualized, were essential for building an authoritarian imperial order. By suppressing multiple points of view from different cultural and philosophical traditions, such centralization achieved two purposes. First, it established universal rules and standards that were themselves the web of the centralized empire. Second, it subjected divergent factions of power, including the emperor's person, to a single system of doctrines, moral values, and social norms. *Huainanzi* represents resistance to this model of totalitarian control of ideas and culture. It proposed a pluralistic model of empire building. Using the cosmology of *Dao* which produced ten thousand things without possessing or dominating them, *Huainanzi*'s vision of the empire was one that allowed space for all cultural traditions and ideas to participate, one that embraced rather than suppressed the great variety of cultural and philosophical traditions, just as nature embraced the great variety of beings.

54 *HSBZ* 6, p. 3b; *HS*, p. 159; for a more detailed account of the success of Confucianism during the Former Han, see Kramers, 1986, pp. 747–66.
55 *HNHLJJ* 13, p. 4b.
56 *HNHLJJ* 13, pp. 4b–5a.

Political Centralization versus Local Autonomy

In the political realm, the symbolic battle over cosmology and cultural centralization had its more visible and tangible manifestation in the blood shed by the local kings, including the king of Huainan, who were defeated in the process of political centralization. While the battle was fought with weapons and human lives, both sides employed symbolic weapons as well – cosmological theories of omens that also had the power to kill.

Dong Zhongshu initiated both the theory of systematic omen interpretation according to the *Spring and Autumn* and the practice of using such interpretations to influence governmental policies. By designating the emperor the single active agent for realizing Heaven's will, Dong also made the emperor responsible for disorders in the universe and in human society, and for catastrophes and strange phenomena that identified and punished human wrongs. It was through interpreting omens that Dong persuaded the Emperor to execute the noble kings.

The most extensive record of Dong's use of omens in politics is his memorial of 140 B.C. According to "Wuxing zhi," the temple and the garden dedicated to the founding emperor Gao caught fire. Dong Zhongshu wrote a memorial to Emperor Wu using an analogy to the four catastrophes of fire recorded in the *Spring and Autumn*, saying that these fires were all messages that Heaven intended to execute the evil powers. This longest memorial recorded in "Wuxing zhi" made an explicit demand on the emperor – to execute his royal relatives with a hardened heart.

Dong first singles out the royal nobility as the contemporary social evil, adding to the historical problems left by Qin and bringing the empire into deep crisis:[57]

> In the past, Qin inherited the malady of the declined Zhou, and failed to cure it. Then Han inherited the malady of Qin and again failed to cure it. . . . On top of that, Han has yet the problem of [the royal] brothers and relatives, who are overbearing, arrogant, extravagant, and indulgent, many of whom are unbridled. That is why we say that this is the "time of deep crisis."

To address such a crisis, Dong appeals to Heaven's authority to speak to the emperor, urging him to execute his evil relatives:[58]

> Heaven descends catastrophes as if speaking to your majesty, saying "facing the present situation, which is full of evil and deep crisis, you cannot govern [the

57 *HSBZ* 27A, pp. 11a–12b; *HS*, pp. 1331–3.

58 Ibid.

empire] except by using great justice and supreme public [standards] (*tai ping zhi gong* 太平至公). Search among the royal relatives and in the kingdoms outside the capital for the ones who have most deviated from the standards, and harden your heart to execute them! Just as I burned down the Temple for Ancestor Gao. . . ." The ones outside the capital that are not proper, even the ones as noble as the Temple of Ancestor Gao, could be burned down, let alone the local kings!

Dong attributes the concept of justice (*ping* 平) and a public standard (*gong* 公) to Heaven's will, using them to compete with the private family relations of the emperor. Dong's proposal was not adopted by Emperor Wu, but rather almost cost him his life. When the memorial was presented to the court, Dong's own disciple, Lü Bushu, who did not know who had written the memorial, told the emperor that the memorial was extremely stupid, and Dong was therefore sentenced to death. Although the emperor pardoned him, Dong dared not interpret bad omens ever again.[59]

But the memorial had long-lasting influence. Eighteen years later, when the king of Huainan was accused of plotting an armed rebellion with some other kingdoms as allies, Emperor Wu remembered the advice of Dong Zhongshu; he appointed Dong's disciple Lü Bushu to take charge of the case of Huainan. Lü used the very text of *Spring and Autumn* as the law, executing the criminals first and reporting to the emperor afterward, and the Emperor always granted his permission. According to *Hanshu*, the second king of Huainan, Liu An, committed suicide, and his family, followers, and the people involved with him were killed by the tens of thousands.[60]

Even though Han historical records excluded the voices of the kings, *Huainanzi*, which survived as part of Emperor Wu's collection, protested against centralization and persecution of the noble kings. It was a bitter protest against the Confucian scholars, for their sowing of discord between the emperor and his blood relatives. *Huainanzi* condemned them for using rites and righteousness to "cause conflicts between the ruler and the subjects, and to generate hatred between flesh and blood."[61]

Disguising their criticism under the cap of the late Xia Dynasty, the authors of *Huainanzi* attribute the contemporary social conflict to evil scholar-officials in the capital who have manipulated the emperor. They condemn the officials' plot to destroy the blood ties within the royal family:[62]

59 *HSBZ* 56, p. 20a; *HS*, p. 2524.
60 *HSBZ* 27Bc, p. 13a; *HS*, p. 1425.
61 *HNHLJJ* 11, p. 1a.
62 *HNHLJJ* 6, pp. 12b–13a.

> The virtuous ruler presides over the throne but does not have peace; the ministers do not speak for *Dao*, and the crowds of subjects please the emperor at the price of corrupting the principles, distancing the emperor's flesh and blood for their self-interests. The evil people gather to plot in secret, getting in between the ruler and subject and between the father and son, competing to please the arrogant ruler and to follow his intentions, causing dissension among people in order to achieve their goals. That is why the ruler and subject are in conflict and do not feel close, and the flesh and blood are distanced and are detached from one another.

This protest also employs catastrophes and strange phenomena to speak to the emperor. According to the text, the social evil caused by evil officials resulted in a series of catastrophes and strange phenomena, including the cracking of temples, the collapse of altars, strange behavior and phenomena in all kinds of animals, the anger of gods, natural catastrophes, and warfare in the human world. After the long list of catastrophes, *Huainanzi* speaks to Emperor Wu, advising him to distance himself from the evil scholar-officials and to use the nonaction way of rule:[63]

> Support the capable, and demote the slanderers and flatterers, silence clever arguments, abandon harsh laws, get rid of the numerous and excessive affairs, block the channel of rumors, close the door to factionalism, abolish knowledge and techniques. . . . therefore to unite with the undivided darkness (*hunming* 混冥, the original state of *Dao*), remove the intentions and release the mind, so that the ten thousand things could return to their own roots.

Ironically, *Huainanzi*, the swan song of Daoist philosophy and a voice of protest against centralization, was presented to Emperor Wu, during whose reign the king of Huainan and many of his followers who participated in the writing of *Huainanzi* lost their lives, Daoism and other non-Confucian schools were banned, and the power of the kingdoms was finally eliminated.

The Emperor as the Pivot of Power

The symbolic struggle over cosmology and emperorship, and the political battle over the centralization of the empire, exemplify the position of the emperor in both the symbolic structure of the cosmos and the political system of the empire. The emperor was not the possessor of political power as an entity, but rather the axis of the cosmos that connected and united horizontal and vertical worlds, the pivot of power at

63 Ibid., pp. 16b–17b.

which political tension was focused, the web of power relations was formed, and different factions of power contested.

The Pivot of Power

Within the power struggle between the scholar-statesmen and the noble kings, the emperor functioned as a pivot of power contestation, through which one political faction interacted with another. After a prolonged struggle, scholar-officials of the central government eventually defeated the noble kings. The triumph was achieved in three major steps, each designed by the scholar-officials of the central government but executed by the emperor. Ban Gu's *Hanshu*, Chapter 14, best summarizes this process:[64]

> Emperor Wen adopted Jia Yi's suggestion and divided the kingdoms of Qi and Zhao. Emperor Jing adopted the strategy of Chao Cuo, and reduced the kingdoms of Wu and Chu. Emperor Wu applied Zhufu Yan's policy, and gave the order of "spreading favor" – ordering the kings to split their kingdoms among their sons, and no son shall be exiled or emigrate [away from the kingdom], and the sons in turn shall divide up their inherited shares among their own sons. From then on, Qi was divided up into seven kingdoms, Zhao into six kingdoms, Liang into five, Huainan into three. . . . Emperor Jing's reign suffered the rebellion of the seven kingdoms; that is why it [the court] weakened the kingdoms by reducing and lowering the officials of the kingdoms. Emperor Wu's reign suffered the rebellion of the kingdoms of Hengshan and Huainan, therefore [the central government] made the law of the "left official" [prohibiting officials who had served the kings to serve in the central government] and punishment for excessive power of the kingdoms. Therefore, the kings only received tax from the kingdom as a source of livelihood, and could not participate in affairs of the government.

According to this description, the first step in reducing the power of the kingdoms was to split large kingdoms into smaller units, which was advocated by Jia Yi, the most influential scholar of Emperor Wen's reign, and implemented by Emperor Wen. The second step was to reduce the territories of the kingdoms and isolate them. This step was designed by Chao Cuo 鼂錯 (200–154 B.C.) – one of the three highest statesmen of the central government, the imperial counselor (*yushi daifu* 御史大夫) – and implemented by Emperor Jing. The third step, the design of a palace counselor (*zhong daifu* 中大夫) named Zhufu Yan 主父偃 (d. ca. 127 B.C.), was to divide up the kingdoms by changing the inheritance law. The new law, promulgated by Emperor Wu, required that all the sons

64 *HSBZ* 14, pp. 3b–4b; *HS*, p. 395.

of a king inherit equal shares of the kingdom. With this law, the kingdoms were quickly fragmented in a few generations. Accompanying these steps, the central government also increased control over the kingdoms by reducing the rank of the officials who served the kingdoms and by appointing the chancellors for the kingdoms from the central government.

In this way, the scholar-statesmen implemented their design for the imperial order through persuasion of the emperor. They reduced the eleven large kingdoms that had existed in 195 B.C. (Map 1) to eighteen isolated, fragmented kingdoms by 108 B.C. (Map 2), with the political power of the kings stripped away. This triumph was not achieved by policy making alone, but was accompanied by bloodshed. The scholars, as "*wen*" 文 as they were, repeatedly persuaded the emperors that for the benefit of the empire and the security of the throne, the emperor must slaughter his own flesh and blood. In a few generations, the eleven kingdoms installed by Emperor Gao had thirteen kings executed as rebels, all brothers and sons of the founding emperor or their descendants.[65] And the people involved in Huainan alone were killed by the tens of thousands.[66]

The noble kings, however, were anything but lambs to be slaughtered. They, too, fought the enemy through their royal relative, the emperor, persuading him that it was the scholar-statesmen who were threatening the throne and the empire. With the help of the emperor, they eliminated their most hated rivals, and the three scholar-statesmen who advocated reducing the kingdoms as just described all lost their lives in the battle against the kings.

The emperor's pivotal position depended on a balance of competing forces. While Emperor Wen did adopt Jia Yi's design of dividing the kingdoms, he also distanced Jia Yi under pressure from Jia Yi's rivals. As a result, Jia Yi never held an important position and committed suicide at the age of thirty-three, depressed over failing to achieve his ambition.[67] Chao Cuo, the second scholar-statesman, did hold one of the three highest offices of the state (as the imperial counselor) and was the favorite statesman of Emperor Jing. Yet he was executed publicly by the same emperor at the insistence of the kings. The story goes that when seven kingdoms revolted in the name of eliminating Chao Cuo in 154 B.C., the emperor had to sacrifice his favorite statesman in the hope of stopping the rebellion and regaining the loyalty of the kings. Even though the rebellion did not end with the execution of Chao Cuo, the

65 *HSBZ* 14, pp. 1a–11b; HS, pp. 391–406.
66 *HSBZ* 27Bc, p. 13a; *HS*, p. 1425.
67 *HSBZ* 48; *HS*, pp. 2221–65.

TUN-HUANG
CHIU-CH'UAN
CHANG-I
WU-WEI
LUNG-HSI
WU-TU
WEN-SHAN
KUANG-HAN
SHU-CHÜN
SHEN-LI
CHIEN-WEI
YÜEH-SUI
TSANG-KO
I-CHOU
CHIAO-CHIH
CHIU-CHEN
JIH-NAN
Commanderies (CHÜN)
A CHIAO-HSI
B KUANG-P'ING
C CHI-YIN
D TA-HO
Kingdoms (KUO)
1 TZU-CH'UAN
2 CHIAO-TUNG
3 CH'ENG-YANG
4 HO-CHIEN
5 CHEN-TING
6 KUANG-CH'UAN
7 CHUNG-SHAN
8 CH'ING-HO

Map 2. The Han Empire, 108 B.C. (Twitchett and Loewe, eds., *The Cambridge History of China*, vol. 1, Cambridge University Press, 1986, pp. 166–7.)

event highlights the point of tension.[68] The third scholar-statesman, Zhufu Yan, the designer of the new law of inheritance, was also executed as a result of charges made by local kings and princes.[69]

The Emperor's Person Divided

As reflected in the struggle between these two political forces, emperorship embodied the tensions among all components of the empire. The emperor himself was often torn between conflicting values, interests, and bonds – between his tie to his "flesh and blood" in the lineage and his dependence on the centralized government, between his imperial consorts and his civil servants, between the military pursuit of conquest and the literary and cultural construction of the empire.

The historical records, all written by the scholar-officials of the central government, repeatedly accused the local kings of being "extremely disrespectful to the emperor" and plotting military rebellion against the throne. But whether the threat from the kingdoms was more to the emperor's person or to the scholars who controlled the bureaucratic state is questionable. Some modern scholars suspect that it was the threat of the kingdom of Huainan as an intellectual and cultural center competing against the Confucianists in the capital, more than its military potential, that caused the armed intervention and the execution of the followers of the king of Huainan.[70]

But while the military and ideological threats have been well analyzed, there was a third dimension of the threat from the kingdoms to the scholars in the capital, that is, the family ties of the kings to the emperor, which often rose above the bureaucratic system. While this dimension has been largely ignored in the scholarship, I argue that, as the pivotal figure of a complex system of power relations, the emperor became the focus of political tensions, and was often torn by conflicting values and interests. The blood tie between the emperor and his lineage, which had a long history as the primary political and social structure itself, was in dramatic conflict with the scholar-officials and the bureaucratic state system. The scholars used the concept of the public (*gong*), the public standard and interests, to fight the private and personal ties of the emperor – his lineage, his consorts, his personal relations. As shown earlier, it was with the argument for the "great justice and supreme public [standards]" that Dong urged Emperor Wu to execute his brothers and uncles. And, opposing the value of the "public," the bond

68 *HSBZ* 49, pp. 22b–24b; *HS*, pp. 2300–2.
69 *HSBZ* 64A, pp. 19a–21a; *HS*, pp. 2802–4.
70 Wallacker, 1972; and Roth, 1992, p. 17.

between "flesh and blood" was what the king of Huainan used in resisting centralization and the elimination of the kingdoms.

The split between the "public" interests of the bureaucratic state symbolized by emperorship and the emperor's own "private" ties with his lineage, consorts, and relations is best demonstrated in the cases of the two kings of Huainan. The first king of Huainan, named Chang 長, was the youngest son of the founding Emperor Gao and the younger brother of Emperor Wen. Chang believed that he was the closest person to the emperor, calling him big brother and often riding in the same carriage with him while hunting. Emperor Wen treated this younger brother in a special manner as well. Ban Gu, the author of *Hanshu*, implies a criticism of Emperor Wen for indulging his brother and not following Han law. Eventually, the chancellor and the imperial counselor, two of the three highest officials of the central government, collected enough charges against Chang to demand that Emperor Wen follow Han law and publicly execute him. The emperor had no heart to do so, and instead, pardoned Chang and spared his life, but deprived him of his title of king. When Emperor Wen later heard that his brother had starved himself to death in protest, he cried with grief. In regret, Emperor Wen went against Jia Yi's advice to diminish the kingdoms; he resumed his dead brother's title of king and made the three sons of Chang kings as well, granting them the territory of the kingdom of Huainan.[71]

The second king of Huainan was one of Chang's three sons, Liu An, who was responsible for the writing of *Huainanzi* and who had close relations with his nephew Emperor Wu. The young Emperor Wu, loving literature and all arts of magic, particularly respected and favored this uncle for his talent and profound knowledge. They had frequent exchanges in poetry and prose. Liu An presented his *Huainanzi* to the emperor, who was said to have loved it and placed it in his imperial collection. They often had dinners together that lasted late into the night, discussing politics and magical techniques, and writing prose and poetry.[72] Liu An's close personal relationship with the emperor, his literary talent and profound knowledge of Daoist techniques, and the strength of his court, which was a center of Daoism and Chu culture that attracted thousands of scholars, together posed a great threat to scholar-officials in the capital, who represented the "public" side of the empire that could not comfortably coexist with the emperor's "private" ties.

Against political rivals such as these brothers and uncles of the emperor, the scholar-statesmen used omens as Heaven's speech to per-

71 *HSBZ* 44, pp. 1a–8a; *HS*, pp. 2135–44. 72 *HSBZ* 44, pp. 8b–9a; *HS*, p. 2145.

suade, or scare, the emperor and to attack their rivals. Even though Dong Zhongshu dared not interpret omens in the court after the events of 140 B.C., his followers continued using omens as persuasion in their memorials to the emperor. In "Wuxing zhi," twenty-five cases of bad omens are interpreted as warnings of rebellions by the kings.

Conclusion

From this reconstruction of the cosmological debate over emperorship and the political struggle between the scholar-bureaucrats and the noble kings, it is apparent that emperorship in Han China was culturally and symbolically produced and reproduced, constantly being contested within cosmological discourse. It was through the cosmological debates over Heaven, *Dao*, and resonance that the emperor's position was argued, the active and nonactive models of sovereignty were contested, centralization and pluralism struggled against one another, and the public standards of the state and the private relations of the emperor competed.[73] In other words, the pivotal position of the emperor in power relations was constantly contested and reproduced through the discursive actions of interpreting omens and theorizing the cosmos.

The emperorship thus discursively produced became the axis of the cosmos and the pivot of the empire – the sole center that vertically connected the human world with Heaven above and Earth below, and horizontally connected the vast web of social and political relations. This pivotal position of emperorship seemingly reclaimed the cosmological and political centrality that the Shang king had occupied in *Sifang* cosmology. But a systematic comparison reveals the fundamental changes that had occurred in the function and nature of rulership as well as in the concept of the center.

The first such change concerns the occupant of the center. The center of the *Sifang* cosmology was occupied by the king together with his entire ruling clan, composed of the deceased ancestors on high and all their living descendants below. The king's primary role was that of connecting the high and low, the living and the dead, by performing the two great services of ancestor worship – ritual sacrifice, which offered flesh and blood to feed the ancestral spirits, and warfare, which provided

73 Michael Loewe has done the most complete study of the changes in the concept of sovereignty from the Qin to the Later Han Dynasty, demonstrating the long process by which the concept of sovereignty based on military and material success was, step by step, replaced by a concept of sovereignty based on ethical ideals and superhuman authorities. See Loewe, 1981; and 1986a, pp. 726–46.

victims to be offered to the ancestors. By such ritual exchange and communication with the ancestors, the king as the highest living descendent of the clan acquired spiritual assistance and communicated with the entire divine world through the ancestors residing there. Such ancestor worship was the foundation of the religious and political domination maintained by the king and the entire ruling clan over all others.

In Han cosmology, by contrast, the axis of the cosmos and pivot of the empire was not a ruling clan, but a single person, the emperor, who claimed his legitimacy not as the descendant of the royal ancestors, but as the single moral exemplary human figure representing the cosmic order in the human world. The emperor received the Mandate of Heaven because of his personal moral quality; he submitted to the moral authority of Heaven, yet functioned as its highest human representative. The Mandate of Heaven embodied an obvious contradiction – that between receiving the mandate of the dynasty on the basis of moral qualifications and inheriting the mandate through lineage. This contradiction persisted throughout imperial history until the last dynasty, the Qing.[74] Despite the fact that inheritance persisted in imperial succession, the emperor's cosmological, ritual, and political roles were defined by his function as the human embodiment of the cosmos of Heaven, rather than as a descendant of a specific lineage. While Han emperors served as the connection between the human world and the divine, just as the Shang king had done, they forged such a connection no longer through communicating with ancestral spirits but through resonating with Heaven by means of moral conduct, imitating the cosmic order in the human world by means of rituals, comprehending Heaven's speech (omens), and realizing Heaven's intention in the human world.

The second change in rulership had to do with the political role of the emperor vis-à-vis other components of power. The king in Shang and Western Zhou was a king among many kings. While a Shang or Zhou king was ranked at the top of the hierarchy, superior to his lords, hereditary ministers, and various ranks of noblility, he shared his authority with the entire ruling clan. At each level of the hierarchy, a lord was a lesser "king" and a minister a lesser "lord"; each performed ritual sacrifices in his own domain as the king did in the capital. The lords and nobility differed from the king in degree but not in kind. Kwang-chih Chang gives an excellent description of such social stratification based on lineage; and Mark Lewis, in studying the last stage of such kingship

74 Zito, 1997, p. 61.

during the Spring and Autumn period, conceptualizes it as "segmented aristocracy."[75]

The emperor of Han, however, was an absolute monarch of all under Heaven. As the axis of the cosmos, he was the only human being who could make the connection between Heaven, Earth, and Man. No one but the emperor could perform sacrificial rituals to Heaven; and the message of Heaven's speech (omens) was delivered exclusively to him. Politically, as the pivot of the empire, the emperor differed from all his subjects in kind rather than in degree. He did not share his authority with his subjects, nor could the subjects reproduce his unique function as human sovereign. *Huainanzi*'s eulogy of the destroyed ties of flesh and blood, of a cosmos in which ten thousand things are equivalent, all sharing the same *Dao*, was but poetic resistance to the formation of such an absolute, supreme monarchy.

The third and most important change in rulership was the very concept of "center" (*zhong* 中) itself. The center in the *Sifang* cosmology was a sacred zone that dominated the peripheries, that was occupied by the royal ancestors and their living descendants and maintained through continued ritual and warfare. Such a ritual and political center was defined cosmologically by *Sifang* and politically by the alien polities called "many *fang*" – both representing the otherness or alienness of the periphery.

The centrality of Han emperorship, by contrast, was not defined by the boundary between "us" and the "others" or between the center and the periphery, but rather by its pivotal function of consolidating all under Heaven into a single unified realm of Man. This unity, however, was anything but a harmonious, well-ordered equilibrium achieved by a "bureaucratic system." Instead, the empire, epitomized by the emperor, embodied intense tensions, conflicts, and constant crisis in a net of entangled power relations. From the pivot of emperorship, the infinite oppositions came into proximity and dynamically interacted with one another. As such a center, emperorship distinguished, divided, yet at the same time connected and mediated these oppositions in the human realm as well as in the cosmos. These oppositions included Heaven versus Man, private versus public, force versus virtue, *Yin* versus *Yang*, civil (*wen*) versus martial (*wu*) modes of rule, high versus low, inner court versus outer court, consort families versus imperial lineage, family versus state, the ruler versus the subject, father versus son, husband versus wife, the senior versus the junior, and so on. All centered on and mediated through the pivot of emperorship, these oppositions formed a complex

75 Kwang-chih Chang, 1983, pp. 9–32; and Lewis, 1990, pp. 28–36, 41.

hierarchy of unequal power relations. Through the same pivot, these unequal relations were brought into dynamic interactions of domination and resistance, of submission and subversion.[76]

What constituted imperial sovereignty, therefore, was not the possession of supreme political power as an entity, but rather the occupation of the pivotal position in a highly complex and dynamic political interaction. If one can say that the pre-imperial power contest was carried out mainly through military confrontations among states, the Han empire ordered competing social forces in a cosmological and political unity. Emperorship functioned as the crucial instrument for such unification. Its position as the center of the cosmos and the pivot of power was constantly reproduced through ceaseless social contest.

The cosmological debate reconstructed in this chapter demonstrates that emperorship, cosmologically constructed and contested, was the pivot of power through which diverse interest groups and rival social forces contested and constrained one another. The Han Empire, having been held as the model of stability, unity, and prosperity for two thousand years, was in fact no more than a body of complex relations of contending fragments of power. The scholars – the "spokesmen" of the empire, to use Inden's term – claimed the grand unity of "all under Heaven" and the absolute power and authority of the emperor, who symbolized the empire. But Chinese people today should no longer be held captive by such a claim.

76 Emperorship as such a pivot was not unique to Han, but rather persisted throughout imperial history. Angela Zito has found a similar pivotal function of the emperor in the eighteenth-century ritual context, in which the emperor functioned as a ritual vessel, was the "centerer" between two oppositions, the "site of transformation between inner emotion and outer cosmos." Her symbolic analysis of portraits of the emperor and ritual texts of the Qing illustrates the pivotal position of the emperor in the ritual context of the Qing, which I find quite similar to emperorship in the political realm of the Han. See Zito, 1997, pp. 30–50.

Conclusion: Cosmology and Power Reconsidered

Throughout Chinese history, the Bronze Age has been seen as the golden age of antiquity, the Han Empire held as the ideal model of imperial order, and the cosmology that emerged during the transition between these two ages acclaimed as the essence of Chinese civilization and the deep structure of the "Chinese mind." This historical study of early China's cosmology demystifies this vision of antiquity, unpacks the imperial model, and debunks the idea of an essential Chinese cultural structure. In doing so, it rethinks culture, power, and the Chinese tradition.

The Mutual Construction of Cosmology and Power

One major theme throughout this study has been that cosmology and political power in early China were mutually constructive. By mutually constructive, I mean that there was no cosmology prior to and independent of its sociopolitical reality, existing as a pure "celestial archetype" or as a deep structure of the Chinese mind that social and political reality imitated or repeated. Cosmology in early China was constructed, and its meaning and structure repeatedly defined, through the power contests of the time. Similarly, there was no given social political structure, independent of and ontologically prior to the cosmology, for which cosmology served as merely a reflection, legitimization, or justification. The power relations of both the Bronze Age and the Han Empire were constructed through a cosmology that defined power relations and prescribed norms for political behavior. The treatment of cosmology as a "discourse" throughout this study has served as a way to describe this mutual and simultaneous production of both the cultural and the political, the symbolic and the material, ideas and practice, signs and power.

This process of mutual and simultaneous construction of the cultural and the political is demonstrated in the conjunction between *Sifang*-center cosmology and the political structure of Shang, a part of what Kwang-chih Chang calls "episodic hegemonies" of the Bronze Age.[1] The hegemonic domination of Shang was conceptually constructed by the cosmology of *Sifang*, which distinguished the peripheral and subordinate "others" – *fang* – from the centrality of the self – the Shang. This conceptual distinction was put into action in ritual performance, warfare, tribute and gift exchange, and the building of temples, tombs, and cities. While *Sifang*-center cosmology structured the political domination of the Shang kings as such, its meaning, structure, and function were also continually being reproduced and redefined by ritual and political action.

The process of mutual construction of cosmology and power further developed during the formation of the empires of Qin and Han. The emerging political forces of this transitional period mobilized correlative cosmologies, *Wuxing* in particular, to destroy the centrality of the old hegemonic state and to construct the new power relations of the imperial era. Using *Wuxing* cosmology, they transformed the static cosmology of *Sifang* by replacing the notion of an eternal center and its surrounding four quarters with five dynamically interacting cosmic phases, conquering and generating one another in sequential order. Besides being used to destroy the old cosmology and political order, *Wuxing* cosmology further functioned as a political discourse of the imperial era, used by diverse factions of power in contesting imperial sovereignty, in defining norms for political behavior, and in constructing emperorship and the power relations that evolved around it.

There was, therefore, no *Wuxing* or *Sifang* cosmology that existed as a pure structure of mind before or outside of the practical, ritual, material, and discursive forms that constituted the total power relations of the society. Nor was there a political system, either the episodic hegemonies or the empire, that was prior to and separate from its particular symbolic basis and cultural expression – that is, cosmology.

This mutual construction of cosmology and power, and more generally of the cultural and the political, is by no means a phenomenon unique to early China or to China as a whole. It is rather a general principle evident also in other times and places, as seen in the "galactic polity" in Southeast Asia as reconstructed by Tambiah,[2] or in the late Roman Empire as shown by Peter Brown.[3] Nevertheless, the precise

1 Kwang-chih Chang, 1986, pp. 305–7.

2 Tambiah, 1976; Tambiah, "The Galactic Polity in Southeast Asia," in his *Culture, Thought, and Social Action*, 1985, pp. 252–86.

3 Brown, 1992.

mode of such mutual construction, and the particular way in which cosmology and power are connected and intertwined, are specific to each society and subject to historical change. As a historical study of how the specific mode of conjunction between cosmology and power changed in early China, this study has attempted to distinguish what continued in Chinese history from what changed.

Shang cosmology constructed the centrality of the Shang king by designating the king's body and his ancestral line as the only channel connecting human beings to the divine world. Thus the king became the embodiment of both divine knowledge of the cosmos and political power on earth. Furthermore, cosmology and political power intertwined with one another mainly through ritual and political action – constant travel, hunting, sacrifice, warfare – and through material construction of temples, tombs, and cities. In other words, this mode of mutual construction was primarily seen in actions and visual forms, resembling what Lévi-Strauss describes as a "concrete" mode of thought, or what Jack Goody depicts as an "oral" mode of communication.[4]

Through the long process of historical transition during the Warring States period and as clearly demonstrated in the Qin and Han Empires, knowledge of the cosmos was separated from the king's political power to become the "cultural capital" of various rising political groups and professionals, eventually being monopolized by the scholar-officials. The means of mutual construction of cosmology and power expanded from concrete performance, action, and material construction to large theoretical system-building, employed in political argument, debate, and persuasion and manifested in the compilation and canonization of texts. Cosmology manifested itself as a political discourse shared by different factions of power, connecting them in a political unity while being used by them in disputes and contestation. While the "concrete," performatory mode of cosmology continued, as seen in the ritual production of Wang Mang, the discursive use of cosmology and its textual form of reproduction also rose to prominence.

The Multifacetedness of Political Culture

The mutual construction of cosmology and power by no means supports the reductionist model of power/ideology reproduction or the model of a dominance/resistance dichotomy.[5] The totality of cosmology and power in early China was a multifaceted phenomenon, having many meanings and functions, and representing many different positions.

4 Lévi-Strauss, (1962) 1966; Goody, (1977) 1987.
5 See the discussion of these two models in the first chapter of the book.

For example, *Wuxing* cosmology had no essence, no substance, no fixed structure or deep meaning. It existed only in its multivalence, in its fluid meaning and ever-changing structure, in the various repetitions of its surface in daily practices. In one context, it was used to represent a physical order of conquering forces of nature and the dynamism of violence underlying cosmic movements. In another context, by contrast, it was used to depict a moral order of birth and nurturing and the moral intentions of Heaven. For the First Emperor of Qin, *Wuxing* was a cosmology of conquest and violence that replaced the moral intentionality of Heaven; while for Han Confucian scholars, such as Liu Xiang and Wang Mang, it was a moral cosmology subordinated to and manifesting the intention of Heaven. In addition to these contradictory usages in political ideology, *Wuxing* was simultaneously a popular cosmology of daily life, used by different layers of society for vastly different purposes.

The multifaceted nature of cosmology is also demonstrated in the fact that even the same interpretation of *Wuxing* cosmology could represent highly dynamic and fluid asymmetrical positions in the politics of domination, rather than representing one fixed binary position such as dominance or resistance. The moral cosmology advocated by scholars of the Former Han, for example, represented resistance to the orthodoxy of imperial sovereignty based on conquest and force, a heritage from the Qin that dominated the court throughout the Former Han. But the same moral cosmology simultaneously served as state ideology, forming the foundation of Han Confucianism, which was crowned as the exclusive philosophy for guiding the government and civil service. This moral cosmology became the capital of scholar-officials, who, resisting the court orthodoxy of a conquest cycle of dynamic succession, used it to assert their dominance over other theories and groups, such as religious specialists and the Daoist philosophy of the king of Huainan.

Deconstructing the "Essence" of Chinese Tradition

Instances of questioning and reevaluating the "essence" of Chinese tradition have marked crucial points of transition in Chinese history. Recent instances include the School of Evidential Research (*kaozheng xue*) of the Qing Dynasty, the May Fourth Movement, and the intellectual movement led by Gu Jiegang and known as Discriminations of Ancient History (*Gushi bian* 古史辯) during the early twentieth century. While the reformulation of Chinese tradition has involved many aspects of its "essence," three fundamental aspects particularly concern this study.

The first is the model of the golden age of antiquity. Gu Jiegang, a revolutionary pioneer in demystifying the "essence" of Chinese culture, suggested that antiquity – the Three Sovereigns, the Five Emperors, and even the first two of the Three Dynasties – was no more than a myth fabricated by Han scholars as political propaganda.[6] Since the beginning of the twentieth century, modern archaeology, together with the discovery of oracle inscriptions of Shang and Zhou, has proven that the Three Dynasties (except the Xia, which is still in need of evidence) not only did exist, but also developed an advanced Bronze Age civilization. While overturning Gu Jiegang's conclusion, modern archaeology nevertheless has brought the demystification of Chinese tradition to a new, higher stage. Instead of denying the historicity of ancient civilization as a later fabrication, modern archaeology has reconfigured the developmental history of ancient China.

After reconstructing the material, social, and political cultures of Bronze Age civilization, for example, Kwang-chih Chang deconstructed the myth of the linear succession of the Three Dynasties as the singular origin of Chinese civilization. Chang concluded that the Three Dynasties "were probably not more than episodic hegemonies of some of the states among many."[7] Thus he replaced the chronological, unilinear succession of the Three Dynasties, the model of dynastic succession held throughout Chinese imperial history, with a model of many coexisting and competing centers of civilization. His conclusion also replaced the idea of a single place of origin of Chinese civilization, the central plain, with multiple origins.

The present work continues demystifying the golden age of antiquity by revealing how the power of the "episodic hegemonies" of the Shang and Western Zhou was culturally constructed through a cosmology of *Sifang*. The myth of a golden antiquity described a succession of three dynasties (Xia, Shang, and Zhou) that manifested the shift of the Mandate of Heaven from the corrupt king of the preceding dynasty to the sage king of the next. This study, however, has revealed that the hegemonic domination of the Shang, and then of the Western Zhou, was expressed in a cosmology of four quarters surrounding a sacred center, from which the king of the ruling clan monopolized the channel of communication with the divine world through ancestor worship, and exerted religious and political dominance over peripheral polities through ritual and warfare.

The second aspect of the "essence" of Chinese tradition is the Han Dynasty, which has been considered the symbol of unity, prosperity,

6 Gu Jiegang, ed., (1926–41) 1986. 7 Kwang-chih Chang, 1986, pp. 305–7.

and strength in imperial China. As Michael Loewe has put it: "The Han dynasty bequeathed to China an ideal and a concept of empire that survived basically intact for two thousand years."[8] Challenging the historical accuracy of this ideal of unity and stability, Loewe has revealed that the Han Dynasty was in fact full of internal conflicts and crises.[9]

The present study enhances this point by revealing the discursive production of and contestation over emperorship. The Han Empire was not an ideal form of bureaucratic state or totalitarian society;[10] it was a political unity only in the sense that it ordered diverse and competing interest groups and social forces into a body of relations. The emperor, the symbol of the empire, was not the possessor of all under Heaven and the source of all power and authority as was long believed, but rather the pivot of power through which diverse compositions of power contested and constrained one another. Emperorship, the pivot of the empire that was culturally constructed and symbolically contested, connected infinite unequal relations into a complex hierarchy and dynamic process of interaction.

The third, and for this study the most relevant, aspect of the reformulation of Chinese tradition concerns cosmology. Gu Jiegang pointed out that "*Wuxing* is the principle of Chinese thought, the belief of the Chinese about the cosmic system; it has had extremely strong and persistent power and influence for over two thousand years."[11] Gu Jiegang and his school of Discriminations of Ancient History gave much attention to undermining this powerful "principle of Chinese thought" by challenging its antique origin and revealing the political motivations of the Han scholars in fabricating this cosmology. Their specific conclusion, that *Wuxing* was invented by politically motivated Qin and Han scholars to justify dynastic succession, has been overturned by the rich body of newly discovered evidence showing that *Wuxing* was used well before the imperial era. Yet their desire to demystify "*Wuxing* cosmology" is still relevant to the rethinking of Chinese culture and tradition in today's context of modernization.

To continue the critical reevaluation of cosmology as "the essence of Chinese civilization" and "the structure of Chinese mind," this book has attempted to deconstruct Chinese cosmology through a comprehensive

8 Loewe, 1986.

9 This aspect of Loewe's scholarship is best represented in Loewe, 1974.

10 Influenced by Max Weber, twentieth-century scholars on China have investigated the origin and development of bureaucracy in early China, describing it as a highly organized, rational, and efficient bureaucratic system that maintained totalitarian control of the society. For representatives of this scholarship, see Balazs, 1964; Bielenstein, 1980; Creel, 1970.

11 Gu Jiegang, (1926–41) 1982, vol. 5, p. 404.

analysis of newly discovered material. It modifies Gu's theory of fabrication by Han scholars, showing that cosmology as an aspect of power had a long history before the Han, as far back as evidence is available, and that the political implications of cosmology went far beyond a few scholars' textual fabrications or interpolations. Cosmology is a cultural phenomenon that is intrinsic to power. As shown in this study, cosmology constituted the power relations of both the "episodic hegemonies" of the Bronze Age and the Han Empire. Cosmology, from *Sifang* to *Wuxing*, was never a pure structure of the Chinese mind, but was always intrinsic to and productive of power, formulating power relations and simultaneously being constructed through power contestation. As an essential component of power, the conception of the cosmic order was mapped out in power struggles, military conquest, political domination, social hierarchies, and violence. As a political discourse, cosmology was the field on which contests over imperial sovereignty, between centralization and pluralism, and between physical force and moral authority were played out.

Therefore I hope to substitute, in the analysis of Chinese history, interactions among diverse social forces for the model of a totalitarian unity, to substitute a pivot of power for an oriental despot, to substitute relationality for the possession of power, and to substitute the political contestation embodied within each cultural-political sphere – whether cosmology, Confucianism, or emperorship – for a presumed eternal essence of these spheres.

Abbreviations

CQFL	*Chunqiu fanlu* 春秋繁露 (Dong Zhongshu 董仲舒, SBCK ed.)
CQFLYZ	*Chunqiu fanlu yizheng* 春秋繁露義證 (Su Yu 蘇輿, prefaced by Wang Xianqian 王先謙, dated 1914)
CYYY	*Bulletin of the Institute of History and Philology* 中央研究院歷史語言研究所集刊 (Taipei: Academia Sinica)
FMT. Rishu	*Fangmatan rishu* 放馬灘日書 (He Shuangquan 何雙全, "Tianshui Fangmatan Qinjian zongshu 天水放馬灘秦簡綜述," *WW* 2 [1989], pp. 23–31)
Heji	*Jiaguwen heji* 甲骨文合集 (Guo Moruo 郭沫若, ed., 13 vols., Beijing: Zhonghua shuju, 1978–83)
HFHD	*The History of the Former Han Dynasty* (Homer H. Dubs, Baltimore: Waverly Press, 1938–55)
HHS	*Hou Hanshu* 後漢書 (Fan Ye 范曄, Beijing: Zhonghua shuju, 1965)
HHSJJ	*Hou Hanshu jijie* 後漢書集解 (Wang Xianqian 王先謙, Changsha, 1915; rpt. in facsimile, Taipei: Yiwen yinshuguan, 1955)
HNZ	*Huainanzi* 淮南子 (Liu An 劉安; references are to Liu Wendian 劉文典)
HNHLJJ	*Huainan honglie jijie* 淮南鴻烈集解 (Liu Wendian 劉文典, Shanghai: Shangwu yinshuguan, 1926; rpt. Taipei: Taiwan Shangwu yinshuguan, 1969)
HS	*Hanshu* 漢書 (Ban Gu 班固, Beijing: Zhonghua shuju, 1962)
HSBZ	*Hanshu buzhu* 漢書補注 (Wang Xianqian 王先謙, Changsha, 1900; rpt. in facsimile, Beijing: Zhonghua shuju, 1983)
Leizuan	*Yinxu jiagu keci leizuan* 殷墟甲骨刻辭類纂, 3 vols. (Yao Xiaosui 姚孝遂 and Xiao Ding 蕭丁, Zhonghua shuju, 1989)
SBBY	*Sibu beiyao* 四部備要

SBCK	*Sibu congkan* 四部叢刊
SHD.Rishu	*Shuihudi Qinmu zhujian* 睡虎地秦墓竹簡 (Shuihudi Qimu zhujian zhengli xiaozu 睡虎地秦墓竹簡整理小組, Beijing: Wenwu chubanshe, 1990)
SJ	*Shiji* 史記 (Sima Qian 司馬遷, Beijing: Zhonghua shuju, 1959)
SJHZ	*Shiki kaichûkôshô* 史記會注考證 (Takigawa Kametarô 瀧川龜太郎, 10 vols., Tokyo: Tôhô bunka gakuin Tôkyô kenkyûjo, 1932–4)
Tun	*Xiaotun nandi jiagu* 小屯南地甲骨 (Zhongguo shehui kexueyuan Kaogu yanjiusuo 中國社會科學院考古研究所, 2 vols. [vol. 1 in 2 parts, vol. 2 in 3 parts], Beijing: Zhonghua shuju, 1980–3)
Ying	*Yingguo suo cang jiagu ji* 英國所藏甲骨集 (2 vols., Sarah Allan 艾蘭, Li Xueqin 李學勤, and Qi Wenxin 齊文心, eds., Beijing: Zhonghua shuju, 1985)
YTL	*Yantielun* 鹽鐵論 (Huan Kuan 桓寬; references are to Wang Liqi 王利器)
YTLJZ	*Yantielun jiaozhu* 鹽鐵論校注 (Wang Liqi 王利器, Shanghai, 1954; rpt. Tianjin: Guji chubanshe, 1983)
YQS.Sun Bin bingfa	*Sun Bin bingfa* 孫臏兵法 (Yinqueshan Hanmu zhujian zhengli xiaozu 銀雀山漢墓竹簡整理小組, Beijing: Wenwu chubanshe, 1975a)
YQS.Sunzi bingfa	*Sunzi bingfa* 孫子兵法 (Yinqueshan Hanmu zhujian zhengli xiaozu 銀雀山漢墓竹簡整理小組, Beijing: Wenwu chubanshe, 1975b)

Works Cited

Akatsuka Kiyoshi 赤塚忠, *Chūgoku kodai no shūkyō to bunka: In ōchō no saishi* 中國古代の宗教と文化：殷王朝の祭祀, Tokyo: Kadokawa shoten, 1977.

Allan, Sarah, *The Shape of the Turtle: Myth, Art, and Cosmos in Early China*, Albany: State University of New York Press, 1991.

Ames, Roger T., *The Art of Rulership*, Honolulu: University of Hawaii Press, 1983.

An Chin-huai, "The Shang City at Cheng-chou," in *Studies of Shang Archaeology*, ed. Kwang-chih Chang, New Haven: Yale Universtiy Press, 1986, pp. 15–48.

An Zuozhang 安作璋, *Ban Gu yu Hanshu* 班固與漢書, Shandong: Renmin chubanshe, 1979.

Anhuisheng wenwu gongzuodui 安徽省文物工作隊, "Fuyang Shuanggudui Xi Han Ruyinhou mu fajue jianbao 阜陽雙古堆西漢汝陰侯墓發掘簡報," *Wenwu* 文物 8 (1978), pp. 14–17, 20, 21.

Appleby, Joyce, Lynn Hunt, and Margaret Jacob, *Telling the Truth about History*, New York: W. W. Norton & Company, 1994.

Arbuckle, Gary, "Inevitable Treason: Dong Zhongshu's Theory of Historical Cycles and Early Attempts to Invalidate the Han Mandate," *Journal of the American Oriental Society* 115 (October/December, 1995), pp. 585–97.

Balazs, Etienne, *Chinese Civilization and Bureaucracy*, ed. A. F. Wright, trans. H. M. Wright, New Haven: Yale University Press, 1964.

Ban Gu 班固, *Hanshu* 漢書, Beijing: Zhonghua Shuju, (1962) 1987 [abbreviation *HS*].

Barmé, Geremie, and Linda Jaivin, eds., *New Ghosts, Old Dreams: Chinese Rebel Voices*, New York: Times Books, 1992.

Barnard, Noel, *The Chu Silk Manuscript*, Canberra: The Australian National University, 1973.

Barnard, Noel, "A New Approach to the Study of Clan-sign Inscriptions of Shang," in *Studies in Shang Archaeology*, ed. Kwang-chih Chang, New Haven: Yale University Press, 1986, pp. 141–206.

Beck, Mansvelt, *The Treatises of Later Han: Their Author, Sources, Contents and Place in Chinese Historiography*, Leiden: E. J. Brill, 1990.

Bielenstein, Hans, *The Bureaucracy of Han Times*, Cambridge: Cambridge University Press, 1980.

Bodde, Derk, "The State and Empire of Ch'in," in *The Cambridge History of China*, vol. 1, ed. Denis Twitchett and Michael Loewe, Cambridge: Cambridge University Press, 1986, pp. 20–102.

Bodde, Derk, *Chinese Thought, Society, and Science*, Honolulu: University of Hawaii Press, 1991.

Bourdieu, Pierre, *Outline of a Theory of Practice*, trans. Richard Nice, Cambridge: Cambridge University Press, 1977.

Bourdieu, Pierre, *Distinction: A Social Critique of the Judgement of Taste*, Cambridge, Mass.: Harvard University Press, 1984.

Brown, Peter, *Power and Persuasion in Late Antiquity: Towards a Christian Empire*, Madison: The University of Wisconsin Press, 1992.

Burke, Peter, *History and Social Theory*, Ithaca: Cornell University Press, 1992.

Chang, Kwang-chih 張光直, "Shangwang miaohao xinkao 商王廟號新考," *CYYY* 15 (1963a), pp. 115–46.

Chang, Kwang-chih 張光直, "Shang Zhou shenhua yu meishu zhong suojian ren yu dongwu guanxi zhi yanbian 商周神話與美術中所見人與動物關係之演變," in *CYYY* 16 (1963b), pp. 115–46.

Chang, Kwang-chih 張光直, "Yinli zhong de erfen xianxiang 殷禮中的二分現象," in *Qingzhu Li Ji xiansheng qishisui lunwen ji* 慶祝李濟先生七十歲論文集, Taipei: Qinghua xuebaoshe, 1965–1967, pp. 353–70.

Chang, Kwang-chih 張光直, "Tan Wang Hai yu Yi Yin de jiri bing zailun Yin Shang wangzhi 談王亥與伊尹的祭日并再論殷商王制," *CYYY* 35 (1973), pp. 111–27.

Chang, Kwang-chih, *Shang Civilization*, New Haven: Yale University Press, 1980a.

Chang, Kwang-chih 張光直, "Shang Zhou guanxi de zai jiantao 商周關係的再檢討," *CYYY* 51 (1980b), pp. 305–10.

Chang, Kwang-chih, *Art, Myth, and Ritual: The Path to Political Authority in Ancient China*, Cambridge, Mass.: Harvard University Press, 1983.

Chang, Kwang-chih 張光直, "Xia Shang Zhou sandai duzhi yu sandai wenhua yitong 夏商周三代都制與三代文化異同," *CYYY* 55.1 (1984), pp. 51–71.

Chang, Kwang-chih, *The Archaeology of Ancient China*, 4th ed., New Haven: Yale University Press, 1986.

Chang, Kwang-chih 張光直, "Shuo Yindai de 'yaxing' 說殷代的亞形," in *Zhongguo qingtong shidai erji* 中國青銅時代貳集, Kwang-chih Chang, Beijing: Sanlian shudian, 1990, pp. 82–94.

Chang Zhengguang 常正光, "Yindai de fangshu yu yinyang wuxing sixiang de jichu 殷代的方術與陰陽五行思想的基礎," *Yinxu buowuyuan yuankan* 殷墟博物苑苑刊 1 (1989), pp. 175–82.

Chang Yuzhi 常玉芝, *Shangdai Zhouji zhidu* 商代周祭制度, Beijing: Zhongguo shehui kexueyuan chubanshe, 1987.

Chen Jiujin 陳久金 and Zhang Jingguo 張敬國, "Hanshan chutu yupian tuxing shikao 含山出土玉片圖形試考," *Wenwu* 文物 4 (1989), pp. 14–7.

Chen Mengjia 陳夢家, "Guwenzi zhong de Shang Zhou jisi 古文字中的商周祭祀," *Yanjing xuebao* 燕京學報 19 (1936), pp. 143–50.

Chen Mengjia 陳夢家, "Xi Zhou tongqi duandai (part 1) 西周銅器斷代（一）," *Kaogu xuebao* 考古學報 9 (1955), pp. 134–75.

Chen Mengjia 陳夢家, *Yinxu buci zongshu* 殷墟卜辭綜述, Beijing: Kexue chubanshe, 1956.

Chen Quanfang 陳全方, *Zhouyuan yu Zhou wenhua* 周原與周文化, Shanghai: Shanghai Renmin chubanshe, 1988.

Cheng, Ann, "Ch'un ch'iu, Kung yang, Ku liang, and Tso chuan," in *Early Chinese Texts: A Bibliographical Guide*, ed. Michael Loewe, Berkeley: Society for the Study of Early China, 1993, pp. 67–76.

Chunqiu fanlu 春秋繁露, *see* Dong Zhongshu 董仲舒.

Comaroff, Jean, *Body of Power, Spirit of Resistance: The Culture and History of a South African People*, Chicago: University of Chicago Press, 1985.

Comaroff, John L., and Jean Comaroff, "The Madman and the Migrant: Work and Labor in the Historical Consciousness of a South African People," *American Ethnologist* 14.2 (1987), pp. 191–209.

Comaroff, John L. and Jean Comaroff, *Of Revelation and Revolution*, vol. 1, Chicago: University of Chicago Press, 1991.

Cook, Constance A. "The Inner and the Outer: Negotiating Space in the Zhou," paper presented at the Association of Asian Studies meeting at Chicago, March, 1997.

Cook, Constance A., "Scribes, Cooks, and Artisans: Breaking Zhou Tradition," *Early China* 20 (1995), pp. 241–78.

Creel, Herrlee G., *Origins of Statecraft in China: The Western Zhou Empire*, Chicago: University of Chicago Press, 1970.

Dittmer, Lowell, and Samuel S. Kim, eds., *China's Quest for National Identity*, Ithaca: Cornell University Press, 1993.

Dirk, Nicholas B., Geoff Eley, and Sherry B. Ortner, eds., *Culture/Power/History: A Reader in Contemporary Social Theory*, Princeton: Princeton University Press, 1994.

Dong Zhongshu 董仲舒, *Chunqiu fanlu* 春秋繁露; references are to Su Yu 蘇輿, *Chunqiu fanlu yizheng* 春秋繁露義證, preface by Wang Xianqian 王先謙, dated 1914, rpt. in facsimile, Taipei: Heluo tushu chubanshe, 1974 [abbreviation *CQFLYZ*].

Dong Zuobin 董作賓, *Yin li pu* 殷曆譜, Li Zhuang: Institute of History and Philology, Academia Sinica, 1945.

Donham, Donald L., *History, Power, Ideology*, Cambridge: Cambridge University Press, 1990.

Dreyfus, Hubert L., and Paul Rabinow, *Michel Foucault: Beyond Structuralism and Hermeneutics*, second edition with an afterword by and an interview with Michel Foucault, Chicago: University of Chicago Press, (1982) 1983.

Dubs, Homer H., *The History of the Former Han Dynasty*, 3 vols., Baltimore: Waverly Press, 1938–55 [abbreviation *HFHD*].

Dubs, Homer H., "The Reliability of Chinese Histories," *The Far Eastern Quarterly* 6.1 (1946), pp. 23–43.

Durkheim, Emile, *The Elementary Forms of the Religious Life*, Joseph Ward Swain, trans., New York: The Free Press, (1915) 1965.

Durkheim, Emile, and Marcel Mauss, *Primitive Classification*, Chicago: University of Chicago Press, 1963.

Eberhard, Wolfram, *Sternkunde und Weltbild im alten China*, Taipei: Chinese Materials and Research Aids Service Center, 1970.

Eberhard, Wolfram, "The Political Function of Astronomy and Astrology in Han China," in *Chinese Thought and Institutions*, ed. John K. Fairbank, Chicago: University of Chicago Press, 1957, pp. 33–70.

Eliade, Mircea, *Le mythe de l'éternel retour: archétypes et répétition*, Paris: Librairie Gallimard, 1949. Translated as *The Myth of the Eternal Return*, London: Routledge & Kegan Paul, 1955.

Eno, Robert, "Was There a High God Ti in Shang Religion?" *Early China* 15 (1990), pp. 1–26.

Falkenhausen, Lothar von, "Issues in Western Zhou Studies: A Review Article," *Early China* 18 (1993), pp. 139–226.

Fan Ye 范曄, *Hou Hanshu* 後漢書, Beijing: Zhonghua shuju, 1982; references are to Wang Xianqian 王先謙, *Hou Hanshu jijie* 後漢書集解, Changsha, 1915; rpt. in facsimile, Taipei: Yiwen yinshuguan, 1955 [abbreviation *HHSJJ*].

Feng Shi 馮時, "Yin buci Sifang feng yanjiu 殷卜辭四方風研究," *Kaogu xuebao* 考古學報 2 (1994), pp. 131–54.

Feng Youlan 馮友蘭, *Zhongguo zhexueshi xinbian* 中國哲學史新編, vols. 2–3, Beijing: Renmin chubanshe, 1983–4.

Foucault, Michel, *Discipline and Punish: The Birth of the Prison*, trans. Alan Sheridan, New York: Vintage Books, (1977) 1979.

Friedrich, Michael, and Michael Lackner, "Once Again: The Concept of Wuxing," *Early China* 9–10 (1983–85), pp. 218–9.

Fu Sheng 伏生, *Shangshu dazhuan* 尚書大傳, SBCK ed.

Fu Juyou 傅舉有 and Chen Songchang 陳松長, *Mawangdui Hanmu wenwu* 馬王堆漢墓文物, Changsha: Hunan chubanshe, 1992.

Gansusheng wenwu kaogu yanjiusuo 甘肅省文物考古研究所, "Gansu Tianshui Fangmatan Zhanguo Qin Han muqun de fajue 甘肅天水放馬灘戰國秦漢墓群的發掘," *Wenwu* 文物 2 (1989), pp. 1–11.

Gao Quxun 高去尋, "Yindai damu de mushi jiqi hanyi zhi tuice 殷代大墓的墓室及其涵義之推測," *CYYY* 39 (1969), pp. 75–188.

Gawlikowski, Krzysztof, and Michael Loewe, "Sun tzu ping fa," in *Early Chinese Texts: A Bibliographical Guide*, ed. Michael Loewe, Berkeley: Society for the Study of Early China, 1993, pp. 446–55.

Geertz, Clifford, *The Interpretation of Cultures*, New York: Basic Books, 1973.

Goody, Jack, *The Domestication of the Savage Mind*, Cambridge: Cambridge University Press, (1977) 1987.

Graham, A. C., *Yin-Yang and the Nature of Correlative Thinking*, Singapore: Institute of East Asian Philosophies, 1986.

Graham, A. C., *Disputers of the Tao*, La Salle, Illinois: Open Court, 1989.

Gramsci, Antoni, *Selections from the Prison Notebooks*, ed. and trans. Quinten Hoare and Geoffrey Nowell Smith, London: Lawrence and Wichart, 1971.

Granet, Marcel, *La Pensée Chinoise*, Paris: Éditions Albin Michel, (1934) 1950.

Gu Jiegang 顧頡剛, "Wude zhongshi xia de zhengzhi he lishi 五德終始下的政治和歷史," *Gushi bian* 古史辨, ed. Gu Jiegang, Shanghai: Shanghai guji chubanshe, (1926–41) 1982, vol.5, pp. 404–616.

Gu Jiegang 顧頡剛, ed., *Gushi bian* 古史辨, 7 vols., Shanghai: Shanghai guji chubanshe, (1926–41) 1982.

Gu Jiegang 顧頡剛, *Shangshu tongjian* 尚書通檢, Beijing: Yen-ching Daxue, Harvard Yen-ching Institute, 1936.

Guanzi 管子, *SBBY* edition.

Guo Moruo 郭沫若, Wen Yiduo 聞一多, and Xu Weiyu 許維遹, eds., *Guanzi jijiao* 管子集校, Beijing: Kexue chubanshe, 1956.

Guo Moruo 郭沫若, ed., Hu Houxuan 胡厚宣, ed. in chief, *Jiaguwen heji* 甲骨文合集, 13 vols., Beijing: Zhonghua shuju, 1978–83.

Guojia wenwuju guwenxian yanjiushi 國家文物局古文獻研究室, ed., *Mawangdui Hanmu boshu* 馬王堆漢墓帛書, vols. 1–4, Beijing: Wenwu chubanshe, 1980–5.

Hansen, Chad, "Chinese Ideographs and Western Ideas," *The Journal of Asian Studies* 52.2 (1993), pp. 373–412.

Harada Masaota 原田正己,"Kan ju no bunshitsu 漢儒の文質説," in *Waseda daigaku tōhō shisō kenkyū shitsu nenpō* 早稲田大學東方思想研究室年報 (Annual report of the Institute of Oriental Thought of Waseda University), vol. 2 (1958), pp. 451–511.

Harper, Donald, "Warring States Natural Philosophy and Occult Thought," in *The Cambridge History of Ancient China*, ed. Michael Loewe and Edward Shaughnessy, Cambridge: Cambridge University Press, 1999, pp. 813–884.

Hayashi Minao 林巳奈夫, "In-chuki ni yurai suru kishin 殷中期に 由來する 鬼神," *Tōhō gakuhō* (Kyoto) 東方學報（京都）41 (1970), pp. 1–70.

Hayashi Minao 林巳奈夫, "In Shū no tenshin 殷周の天神," *Koshi shunjū* 古史春秋 6 (1989), pp. 2–25.

He Shuangquan 何雙全, "Tianshui Fangmatan Qinjian zongshu 天水放馬灘秦簡綜述," *Wenwu* 文物 2 (1989), pp. 23–31.

Hebdige, Dick, *Subculture: The Meaning of Style*, New York: Methuen, 1979.

Henderson, John B., *The Development and Decline of Chinese Cosmology*, New York: Columbia University Press, 1984.

"Hongfan 洪範"; references are to Kong Yingda 孔穎達, *Shangshu zhengyi* 尚書正義 12, in *Shisanjing zhushu* 十三經注疏, vol. 1, ed. Ruan Yuan 阮元, rpt. in facsimile, Beijing: Zhonghua shuju, 1980.

Hsiao, Kung-chuan, *A History of Chinese Political Thought*, vol. 1, trans. F. W. Mote, Princeton: Princeton University Press, 1979.

Hsu Cho-yun and Katheryn M. Linduff, *Western Zhou Civilization*, New Haven: Yale University Press, 1988.

Hu Houxuan 胡厚宣, *Jiaguxue Shangshi luncong chuji* 甲骨學商史論叢初集, Chengdu: Qilu daxue, 1944.

Hu Houxuan 胡厚宣, "Gongfang kao 𢀛方考," in *Jiaguxue Shangshi luncong chuji* 甲骨學商史論叢初集, vol. 1, Chengdu: Qilu daxue, 1944a.

Hu Houxuan 胡厚宣, "Jiaguwen Sifang fengming kaozheng 甲骨文四方風名考證,"

in *Jiaguxue Shangshi luncong chuji* 甲骨學商史論叢初集, vol. 1, Chengdu: Qilu daxue, 1944b.

Hu Houxuan 胡厚宣, "Lun wufang guannian ji 'Zhongguo' chengwei zhi qiyuan 論五方觀念及中國稱謂之起源," in *Jiaguxue Shangshi luncong chuji* 甲骨學商史論叢初集, vol. 1, Chengdu: Qilu daxue, 1944c.

Hu Houxuan, "Yindai zhi tianshen chongbai 殷代之天神崇拜," in *Jiaguxue Shangshi luncong chuji* 甲骨學商史論叢初集, vol. 1, Chengdu: Qilu daxue, 1944d.

Hu Houxuan 胡厚宣, "Shi Yindai qiu nian yu Sifang he Sifang feng de jisi 釋殷代求年於四方和四方風的祭祀," *Fudan xuebao (Renwen kexue)* 復旦學報 (人文科學) 1 (1956), pp. 49–86.

Hu Houxuan 胡厚宣, "Yin buci zhong de shangdi he wangdi 殷卜辭中的上帝和王帝," *Lishi yanjiu* 歷史研究 9–10 (1959), pp. 23–50, 89–110.

Hu Houxuan 胡厚宣, ed., *Jiaguwen yu Shang Yin shi* 甲骨文與商殷史, Shanghai: Shanghai guji chubanshe, 1983.

Huainanzi 淮南子, *see* Liu An 劉安.

Huan Kuan 桓寬, *Yantielun* 鹽鐵論; references are to Wang Liqi 王利器, *Yantielun jiaozhu* 鹽鐵論校注, Shanghai, 1954; rpt. Tianjin: Guji chubanshe, 1983 [abbreviation *YTLJZ*].

Huangdi neijing suwen 黃帝內經素問, Beijing: Renmin weisheng chubanshe, 1963.

Hulsewé, A. F. P., "The Two Early Han I Ching Specialists Called Ching Fang," *T'oung Pao* 72 (1986), pp. 161–2.

Hulsewé, A. F. P., "Hanshu," in *Early Chinese Texts: A Bibliographical Guide*, ed. Michael Loewe, Berkeley: Society for the Study of Early China, 1993, pp. 129–36.

Hulsewé, A. F. P., "Notes on the Historiography of the Han Period," *Historians of China and Japan*, ed. W. G. Beasley and E. G. Pulleyblank, London: Oxford University Press, 1961, pp. 31–43.

Ikeda Suetoshi 池田末利, *Chūgoku kodai shūkyōshi kenkyū: Seido to shisō* 中國古代宗教史研究：制度と思想, Tokyo: Tokai University Press, 1981.

Inden, Ronald, *Imagining India*, Cambridge: Basil Blackwell, 1990.

Itō Michiharu 伊藤道治, *Chūgoku kodai ōchō no keisei* 中國古代王朝の形成, Tokyo: Sōbunsha, 1975.

Ito Michiharu 伊藤道治, "Shū Buō to Rakuyū–kason mei to Itsushūsho doyū 周武王と雒邑一珂尊銘と逸周書度邑," in *Uchida Ginpū hakushi shōju kinen Tōyōshi ronshū* 內田吟風博士頌壽紀念東洋史論集, Kyoto: Dohosha, 1978, pp. 41–53.

Jia Yi 賈誼, *Xinyu* 新語, SBBY edition.

Jiang Liangfu 姜亮夫, "San Chu suochuan gushi yu Qi, Lu, san Jin yitong 三楚所傳古式與齊，魯，三晉異同," *Lishi yanjiu* 歷史研究 4 (1979).

Jingmenshi bowuguan 荊門市博物館, *Guodian Chumu zhujian* 郭店楚墓竹簡, Beijing: Wenwu chubanshe, 1998.

Jingzhou diqu bowuguan 荊州地區博物館, "Jiangling Wangjiatai 15 hao Qinmu 江陵王家台15號秦墓," *Wenwu* 1 (1995), pp. 1–37, 42.

Kalinowski, Marc, "Les Traités de Shuihudi et l'Hémérologie Chinoise à la Fin des Royaumes-Combattants," *T'oung Pao* 72 (1986), pp. 175–228.

Kalinowski, Marc, *Cosmologie et Divination dans la Chine Ancienne: le Compendium des Cinq Agents,* Paris: Ecole Française d'Extrême-Orient, 1991.

Kantorowicz, Ernst H., *The King's Two Bodies: A Study in Mediaeval Political Theology,* Princeton: Princeton University Press, 1957.

Keightley, David N., "The Religious Commitment: Shang Theology and the Genesis of Chinese Political Culture," *History of Religions* 17 (1978a), pp. 211–225.

Keightley, David N., *Sources of Shang History: The Oracle-Bone Inscriptions of Bronze Age China,* Berkeley: University of California Press, 1978b.

Keightley, David N., "The Late Shang State: When, Where, and What?" in *The Origins of Chinese Civilization,* ed. Keightley, Berkeley: University of California Press, 1983, pp. 523–64.

Keightley, David N., "Late Shang Divination: The Magico-Religious Legacy," in *Explorations in Early Chinese Cosmology,* ed. Henry Rosemont, Jr., Journal of the American Academy of Religion Studies, Thematic Studies no. 50.2, Chico, California: Scholars Press, 1984, pp. 11–34.

Keightley, David N., "Shang Divination and Metaphysics," *Philosophy East and West* 38.4 (October, 1988), pp. 367–97.

Keightley, David N., "The Quest for Eternity in Ancient China: The Dead, Their Gifts, Their Names," in *Ancient Mortuary Traditions of China: Papers on Chinese Ceramic Funerary Scriptures,* ed. George Kuwayama, Los Angeles: Los Angeles County Museum of Art, 1991, pp. 12–25.

Keightley, David N., "Chinese Religions: The State of the Field: Part I, Early Religous Traditions: The Neolithic Period through the Han Dynasty (ca. 4000 B.C.E.–220 C.E.): Neolithic and Shang Periods," *Journal of Asian Studies* 54.1 (February, 1995), pp. 128–45.

Keightley, David N., "The Shang: China's First Historical Dynasty," in *The Cambridge History of Ancient China,* ed. Michael Loewe and Edward Shaughnessy, Cambridge: Cambridge University Press, 1999, pp. 232–291.

Keightley, David N., *The Ancestral Landscape: Time, Space, and Community in Late Shang China (ca. 1200–1045 B.C.),* Berkeley: Institute of East Asian Studies, University of California, forthcoming.

Kramers, Robert P., "The Development of the Confucian Schools," in *The Cambridge History of China,* ed. Denis Twitchett and Michael Loewe, Cambridge: Cambridge University Press, 1986, vol. 1, pp. 747–66.

Kunst, Richard A., "More on Xiu and Wuxing, with an Addendum on the Use of Archaic Construction," *Early China* 3 (1977), pp. 67–9.

Le Blanc, Charles, *Huai-nan tzu: Philosophical Synthesis in Early Han Thought,* Hong Kong: Hong Kong University Press, 1985.

Legge, James L., trans., *The Chinese Classics,* vol. 3: *The Shoo King,* rpt. Hong Kong: Hong Kong University Press, 1960.

Legge, James L., trans., *The Chinese Classics*, vol. 4: *The She King*, rpt. Hong Kong: Hong Kong University Press, 1960.

Legge, James L., trans., *The Chinese Classics*, vol. 5: *The Ch'un ts'ew with the Tso chuan*, rpt., Hong Kong: Hong Kong University Press, 1960.

Lévi-Strauss, Claude, *The Savage Mind*, Chicago: University of Chicago Press, (1962) 1966.

Lewis, Mark Edward, *Sanctioned Violence in Early China*, Albany: State University of New York Press, 1990.

Li Hansan 李漢三, *Xian Qin Liang Han Yin-Yang Wuxing xueshuo* 先秦兩漢陰陽五行學說, Taipei: Weixin shuju, 1981.

Li Ling 李零, *Changsha Zidanku Zhanguo Chu boshu yanjiu* 長沙子彈庫戰國楚帛書研究, Beijing: Zhonghua Shujiu, 1985.

Li Ling 李零, "'Shi tu' yu Zhongguo gudai de yuzhou moshi 式圖與中國古代的宇宙模式," *Jiuzhou xuekan* 九州學刊 4.1 (1991a), pp. 5–53; 4.2 (1991b), 49–76.

Li Ling 李零, *Zhongguo fangshu kao* 中國方術考, Beijing: Renmin Zhongguo chubanshe, 1993.

Li Xiaoding 李孝定, *Jiagu wenzi jishi* 甲骨文字集釋, Zhongyang yanjiuyuan lishi yuyan yanjiusuo zhuankan zhi wushi 中央研究院歷史語言研究所專刊之五十, 8 vols., Nangang, 1965.

Li Xueqin 李學勤, "Shangdai de sifeng yu sishi 商代的四風與四時," *Zhongzhou xuekan* 中州學刊 5 (1985), pp. 99–101.

Li Xueqin, "A Neolithic Jade Plaque and Ancient Chinese Cosmology," *National Palace Museum Bulletin* 27.5–6, (November-December 1992, January-Feburary 1993), pp. 1–8.

Li Xueqin 李學勤, "Jingmen Guodian Chujian zhong de 'Zi Si zi'" 荊門郭店楚簡中的子思子, *Wenwu tiandi* 文物天地 2 (1998), pp. 28–30.

Lian Shaoming 連劭名, "Shangdai de Sifang fengming yu bagua 商代的四方風名與八卦," *Wenwu* 文物 11 (1988), pp. 40–4.

Lian Shaoming 連劭名, "Shangdai de qiudong sidian 商代的秋冬祀典," *Renwen Zhongguo xuebao* 人文中國學報 3 (1996), pp. 165–78.

Liang Qichao 梁啟超, "Yin-Yang Wuxing shuo zhi laili 陰陽五行說之來歷," in *Gushi bian* 古史辨, 7 vols., ed. Gu Jiegang, Shanghai: Shanghai guji chubanshe, (1924–41) 1982, vol. 5, pp. 343–62.

Liang Siyong 梁思永 and Gao Quxun 高去尋, *Houjiazhuang* 侯家莊, 7 vols. (tomb no. 1001, parts a–b, 1962; no. 1002, 1965; no. 1003, 1967; no. 1217, 1968; no. 1004, 1970; no. 1550, 1974; no. 1550, 1976), Taipei: Institute of History and Philology, 1962–76.

Lin Jianming 林劍鳴, "Rishu yu Qin Han shidai de lizhi 日書與秦漢時代的吏制," *Xinshixue* 新史學 2.2 (1991), pp. 31–51.

Lin Yun 林沄, "Tian Wang gui si yu tian shi xin jie 天亡簋祀于天室新解," paper presented at the international conference on Western Zhou civilization, Xi'an, Shaanxi, July 1993.

Liu An 劉安, *Huainanzi* 淮南子; references are to Liu Wendian 劉文典, *Huainan honglie jijie* 淮南鴻烈集, Shanghai: Shangwu yinshuguan, 1926; rpt. Taipei: Taiwan Shangwu yinshuguan, 1969 [abbreviation *HNHLJJ*].

Liu Guosheng 劉國勝, "Shashi chutu 'shizhan' zhujian ji qi xianguan wenti yanjiu 沙市出土'式占'竹簡及其相關問題研究," M.A. thesis, Wuhan University, 1995.
Liu Jie 劉節, *Zhongguo gudian zongzu yizhi shulun* 中國古代宗族移植史論, Taipei: Zhengzhong shuju, 1948
Liu Wendian 劉文典, *see* Liu An 劉安.
Liu Zhiji 劉知己, *Shitong* 史通; references are to Pu Qilong 浦起龍 (fl. 1730–1752), *Shitong tongshi* 史通通釋, 2 vols., rpt. Shanghai: Shanghai guji Chubanshe, 1978.
Loewe, Michael, *Crisis and Conflict in Han China: 104 B.C. to 9 A.D.*, London: George Allen & Unwin, 1974.
Loewe, Michael, "Water, Earth and Fire – the Symbols of the Han Dynasty," *Nachrichten* 125 (1979a), pp. 63–8; rpt. in *Divination, Mythology, and Monarchy in Han China*, Cambridge: Cambridge University Press, 1994, pp. 55–60.
Loewe, Michael, *Ways to Paradise*, London: George Allen & Unwin, 1979b.
Loewe, Michael, "The Authority of the Emperors of Ch'in and Han," in *State and Law in East Asia*, eds. Dieter Eikemeier and Herbert Franke, Wiesbaden: O. Harrassowitz, 1981.
Loewe, Michael, *Chinese Ideas of Life and Death*, London: George Allen & Unwin, 1982.
Loewe, Michael, "The Term K'an-yü and the Choice of Moment," *Early China* 9/10 (1983–5), pp. 204–17; rpt. in *Divination, Mythology and Monarchy*, Cambridge: Cambridge University Press, 1994, pp. 112–20.
Loewe, Michael, "The Concept of Sovereignty," in *The Cambridge History of China*, ed. Denis Twitchett and Michael Loewe, Cambridge: Cambridge University Press, 1986a, vol. 1, pp. 726–742.
Loewe, Michael, "The Former Han Dynasty," in *The Cambridge History of China*, ed. Denis Twitchett and Michael Loewe, Cambridge: Cambridge University Press, 1986b, vol. 1, pp. 103–222.
Loewe, Michael, "The Religious and Intellectual Background," in *The Cambridge History of China*, ed. Denis Twitchett and Michael Loewe, Cambridge: Cambridge University Press, 1986c, vol. 1, pp. 649–725.
Loewe, Michael, "The Structure and Practice of Government," in *The Cambridge History of China*, ed. Denis Twitchett and Michael Loewe, Cambridge: Cambridge University Press, 1986d, vol. 1, pp. 463–490.
Loewe, Michael, "The Almanacs (Jih-shu) from Shui-hu-di: A Preliminary Survey," *Asia Major*, 3rd series, vol. 1, part 2 (1988), pp. 1–27; rpt. in *Divination, Mythology and Monarchy in Han China*, Cambridge: Cambridge University Press, 1994, pp. 214–35.
Loewe, Michael, ed., *Early Chinese Texts: A Bibliographical Guide*, Berkeley: Society for the Study of Early China, 1993.
Loewe, Michael, *Divination, Mythology and Monarchy in Han China*, Cambridge: Cambridge University Press, 1994.
Loewe, Michael, "Huang-Lao Thought and the *Huainanzi*: A Review Article," *Journal of the Royal Asian Society* 4.3 (1995), pp. 377–95.
Loewe, Michael, and Edward L. Shaughnessy, eds., *The Cambridge History of*

Ancient China: From the Origins of Civilization to 221 B.C., Cambridge: Cambridge University Press, 1999.

Luo Zhenyu 羅振玉, *Yinxu shuqi kaoshi* 殷墟書契考釋, n.p., 1915.

Luo Zhenyu 羅振玉, *Sandai jijin wencun* 三代吉金文存, Pojuezhai, 1936.

Lü Simian 呂思勉, *Qin Han shi* 秦漢史, Shanghai: Shanghai guji chubanshe, (1947) 1986.

Lüshi chunqiu 呂氏春秋, *SBBY* edition [abbreviation *LSCQ*].

Ma Chengyuan 馬承源, ed., *Shang Zhou qingtongqi mingwen xuan* 商周青銅器銘文選, 4 vols., Beijing: Wenwu chubanshe, 1988.

Major, John S., "A Note on the Translation of Two Technical Terms in Chinese Science: Wuxing and Xiu," *Early China* 2 (1976), pp. 1–3.

Major, John S., "Reply to Richard Kunst's Comments on Xiu and Wuxing," *Early China* 3 (1977), pp. 69–70.

Major, John S., "The Five Phases, Magic Squares, and Schematic Cosmography," in *Exploration in Early Chinese Cosmology*, ed. Henry Rosemcord, Jr., Journal of American Academy of Religion Studies, Thematic Studies no. 50.2, Chico, California: Scholars Press, 1984, pp. 133–66.

Major, John S., "The Meaning of Hsing-te," in *Chinese Ideas about Nature and Society*, ed. Charles le Blanc and Susan Blader, Hong Kong: Hong Kong University Press, 1987, pp. 281–92.

Major, John S., *Heaven and Earth in Early Han Thought, Chapters Three, Four and Five of the Huainanzi*, Albany: State University of New York Press, 1993.

Marx, Karl, and Friedrich Engels, *The German Ideology*, London: Lawrence and Wishart, (1845–6) 1965.

Marx, Karl, and Friedrich Engels, *Selected Works*, 2 vols., Moscow, 1958.

Moore, Sally F., *Social Facts and Fabrications: "Customary" Law on Kilimanjaro, 1880–1980*, Cambridge: Cambridge University Press, 1986.

Moore, Sally F., ed., *Moralizing States and the Ethnography of the Present*, American Ethnological Society Monograph Series, no. 5, 1993.

Mozi 墨子; references are to Sun Yirang 孫詒讓 (1848–1908), *Mozi jiangu* 墨子閒詁, in *Zhuzi jicheng* 諸子集成, vol. 4, 1954; rpt. Beijing: Zhonghua Shuju, 1986.

Needham, Joseph, and Wang Ling, *Science and Civilisation in China*, vol. 2, Cambridge: Cambridge University Press, 1956.

Needham, Joseph, and Wang Ling, *Science and Civilization in China*, vol. 3, Cambridge: Cambridge University Press, 1959.

Nivison, David, "A New Study of Xiaotun Yinxu Wenzi Jiabian 2416," paper presented to the international conference on Shang culture, Anyang, P.R.C., September, 1987.

Nivison, David, "'Question' Questions," *Early China* 14 (1989), pp. 115–26.

Nylan, Michael, *The Shifting Center: The Original "Great Plan" and Later Readings*, Monumenta Serica Monograph Series, no. 24, 1992.

Onselen, Charles van, *Chibaro: African Mine Labour in Southern Rhoedisia 1900–1933*, London: Pluto Press, 1976.

Pang Pu 龐樸, "Mawangdui boshu jiekai le Si Meng Wuxing zhi mi 馬王堆帛書解開了思孟五行之迷," in *Mawangdui Hanmu yanjiu* 馬王堆漢墓研究, ed. Hunansheng bowuguan 湖南省博物館, Changsha: Hunan Renmin chubanshe, 1981, pp. 119–32.

Pankenier, David W., "Astronomical Dates in Shang and Western Zhou," *Early China* 7 (1981–2), pp. 2–37.

Pankenier, David W., "Mozi and the Dates of Xia, Shang, and Zhou: A Research Note," *Early China* 9–10 (1983–5), pp. 175–83.

Pankenier, David W., "The Bamboo Annals Revisited: Problems of Method in Using the Chronicle as a Source for the Chronology of Early Zhou, Part 2: The Congruent Mandate Chronology in Yi Zhou shu," *Bulletin of the School of Oriental and African Studies* 55.3 (1992), pp. 498–510.

Pankenier, David W. "Cosmo-Political Background of Heaven's Mandate," *Early China* 20 (1995), pp. 121–176.

Poo Mu-chou 蒲慕州, "Shuihudi Qinjian rishu de shijie 睡虎地秦簡日書的世界," *CYYY* 62.4 (1992), pp. 623–675.

Poo Mu-chou 蒲慕州, *In Search of Personal Welfare: A View of Ancient Chinese Religion,* Albany: State University Press, 1998.

Price, Nancy T., "The Privot: Comparative Perspectives from the Four Quarters," *Early China* 20 (1995), pp. 93–120.

Qian Mu 錢穆, *Liang Han jingxue jinguwen pingyi* 兩漢經學今古文評議, Hong Kong: Xinya yanjiusuo, 1958.

Qian Mu 錢穆, *Xian Qin zhuzi xinian* 先秦諸子繫年, 2 vols., 1935, rpt. Beijing: Xinhua shudian, 1985.

Qin jian zhengli xiaozu 秦簡整理小組, *Qin Han jiandu lunwenji* 秦漢簡牘論文集, Lanzhou: Gansu Renmin chubanshe, 1989.

Qiu Xigui, "An Examination of Whether the Charges in Shang Oracle-Bone Inscriptions are Questions," *Early China* 14 (1989), pp. 77–114.

Queen, Sarah, *From Chronicle to Canon: The Hermeneutics of the Spring and Autumn Annals according to Dong Zhongshu,* Cambridge: Cambridge University Press, 1996.

Ran Zhaode 冉昭德, "Ban Gu yu Hanshu 班固與漢書," *Lishi jiaoxue* 歷史教學 4 (1962), pp. 33–8.

Rao Zongyi 饒宗頤 and Zeng Xiantong 曾憲通, *Yunmeng Qinjian rishu yanjiu* 雲夢秦簡日書研究, Hong Kong: Chinese University Press, 1982.

Rao Zongyi 饒宗頤 and Zeng Xiantong 曾憲通, *Chu boshu* 楚帛書, Hong Kong: Zhonghua shuju, 1985.

Richett, W. Allyn, trans., *Guanzi: Political, Economic, and Philosophical Essays from Early China,* Princeton: Princeton University Press, 1985.

Rosemont, Henry, Jr., ed., *Exploration in Early Chinese Cosmology,* Journal of American Academy of Religion Studies, Thematic Studies no. 50.2, Chico, California: Scholars Press, 1984.

Roth, Harold D., *The Textual History of the Huai-nan-tzu,* Association of Asian Studies Monograph Series, 1992.

Ruan Yuan 阮元, *Shisanjing zhushu* 十三經注疏, rpt. in facsimile, Beijing: Zhonghua shuju, 1980.

Sahlins, Marshall, *Historical Metaphors and Mythical Realities: Structure in the Early History of the Sandwich Islands Kingdom,* Association for the Study of Anthropology in Oceania, Special Publication no. 1, Ann Arbor: University of Michigan Press, 1981.

Sahlins, Marshall, *Islands of History*, Chicago: University of Chicago Press, 1985.

Savage, William E., "Archetypes, Model Emulation, and the Confucian Gentleman," *Early China* 17 (1992), pp. 1–25.

Schuessler, Axel, *A Dictionary of Early Zhou Chinese,* Honolulu: University of Hawaii Press, 1987.

Schwartz, Benjamin I., *The World of Thought in Ancient China,* Cambridge, Mass.: Harvard University Press, 1985.

Scott, James C., *The Weapons of the Weak: Everyday Forms of Peasant Resistance,* New Haven: Yale University Press, 1985.

Shanxisheng wenwu gongzuodui weiyuanhui 山西省文物工作隊委員會, *Houma mengshu* 侯馬盟書, Shanghai: Wenwu chubanshe, 1976.

Shangshu 尚書; references are to Kong Yingda 孔穎達, *Shangshu zhengyi* 尚書正義, in *Shisanjing zhushu* 十三經注疏, vol. 1, ed. Ruan Yuan 阮元, rpt. in facsimile, Beijing: Zhonghua shuju, 1980.

Shaughnessy, Edward L., *Sources of Western Zhou History,* Berkeley and Los Angeles: University of California Press, 1992.

Shaughnessy, Edward L. 夏含夷, "Yanguo tongqi zukao chenghao yu Zhouren shifa de qiyuan 燕國銅器祖考稱號與周人諡法的起源," in *Beijing jiancheng 3040 nian ji Yan wenming guoji xueshu yantao hui huiyi zhuanji* 北京建城3040年及燕文明國際學術研討會會議專集, Beijing: Yanshan chubanshe, 1997, pp. 320–26.

Shaughnessy, Edward L., "Western Zhou History," in *The Cambridge History of Ancient China,* ed. Michael Loewe and Edward Shaughnessy, Cambridge: Cambridge University Press, 1999, pp. 209–351.

Shen Jianhua 沈建華, "Cong jiaguwen guizi kan Yindai yili zhong de Wuxing guannian qiyuan 從甲骨文圭字看殷代儀禮中的五行觀念," *Wenwu* 文物 5 (1993), pp. 77–79, 96.

Shima Kunio 島邦男, *Gugyo shisō to Raiki tetsurei no kenkyū* 五行思想と禮記月令の研究, Tokyo, Kyūko shoin, 1971.

Shima Kunio 島邦男, *Inkyo bokuji kenkyū* 殷墟卜辭研究, Hirosaki: Hirosaki daigaku Chūgoku gaku kenkyūkai, 1958.

Shirakawa Shizuka 白川靜, *Kinbun tsūshaku* 金文通釋, 56 vols. to date, Kobe: Hakutsuru bijutsukanshi, 1962–84.

Shuihudi Qimu zhujian zhengli xiaozu 睡虎地秦墓竹簡整理小組, *Shuihudi Qinmu zhujian* 睡虎地秦墓竹簡, Beijing: Wenwu chubanshe, 1990.

Sima Qian 司馬遷, *Shiji* 史記, Beijing: Zhonghua shuju, 1959.

Sivin, Nathan, *Traditional Medicine in Contemporary China,* Ann Arbor: Center for Chinese Studies, University of Michigan, 1987.

Sivin, Nathan, "Huangti neiching," in *Early Chinese Texts: A Bibliographical Guide,* ed. Michael Loewe, Berkeley: Society for the Study of Early China, 1993, pp. 196–215.

Steinhardt, Nancy N., *Chinese Imperial City Planning*, Honolulu: University of Hawaii Press, 1990.

Su Xiaokang 蘇曉康 and Wang Luxing 王魯祥, eds., *Heshang* 河殤, Beijing: Xiandai chubanshe, 1988.

Su Yu 蘇輿, *see* Dong Zhongshu.

Takigawa Kametarō 瀧川龜太郎, *Shiki kaichū kōshō* 史記會注考證, 10 vols., Tokyo: Tōhō bunka gakuin Tōkyō kenkyūjo, 1932–4.

Tambiah, Stanley J., *World Conqueror and World Renouncer: A Study of Religion and Polity in Thailand against a Historical Background*, Cambridge: Cambridge University Press, 1976.

Tambiah, Stanley J., *Culture, Thought, and Social Action*, Cambridge, Mass.: Harvard University Press, 1985.

Tambiah, Stanley J., *The Buddhist Conception of Universal King and Its Manifestations in South and Southeast Asia*, Kuala Lumpur: University of Malaya, 1987.

Tang Jinyu 唐金裕, "Xi'an xijiao Handai jianzhu yizhi fajue baogao 西安西郊漢代建築遺址發掘報告," *Kaogu xuebao* 考古學報 2 (1959), pp. 45–54, plates 1–9.

Thompson, Edward Palmer, *The Making of the English Working Class*, London: V. Collacz, 1963.

Tjan Tjoe Som, *Bo Hu T'ung*, 2 vols, Leiden: E. J. Brill, 1949.

Tu, Cheng-sheng 杜正勝, *Gudai shehui* 古代社會, Taipei: Yunzhen wenhua shiye kufen youxian gongsi, 1992.

Tu, Wei-ming, "The Continuity of Being," in *Confucian Thought: Selfhood as Creative Transformation*, ed. Wei-ming Tu, Albany: State University of New York Press, 1985, pp. 35–42.

Tu, Wei-ming, ed., *The Living Tree: The Changing Meaning of Being Chinese Today*, Stanford: Stanford University Press, 1991.

Twitchett, Denis, and Michael Loewe, eds., *The Cambridge History of China*, vol. 1, Cambridge: Cambridge University Press, 1986.

Wallacker, Benjamin, "Liu An, Second King of Huai-nan (180?–122 B.C.)," *Journal of the American Oriental Society* 92 (1972), pp. 36–51.

Wang Guowei 王國維, "Mingtang miaoqin tongkao 明堂廟寢通考," in *Guantang jilin* 觀堂集林, vol. 3, rpt. Beijing: Zhonghua shuju, (1921) 1984.

Wang Guowei 王國維, "Yin Zhou zhidu lun 殷周制度論," in *Guantang jilin* 觀堂集林, vol. 10, rpt. Beijing: Zhonghua shuju, (1921) 1984.

Wang Liqi 王利器, *see* Huan Kuan 桓寬.

Wang Tao, "Color Symbolism in Late Shang China," Ph.D. dissertation, University of London, 1993a.

Wang Tao, "Shang Color Symbolism and the Development of the Wuxing Theory", paper presented at the Association of Asian Studies annual meeting, 1993b.

Wang Xianqian 王先謙, *Hou Hanshu jijie* 後漢書集解, Changsha, 1915; rpt. in facsimile, Taipei: Yiwen chubanshe, 1955 [abbreviation *HHSJJ*].

Wang Xianqian 王先謙, *Hanshu buzhu* 漢書補注, Changsha, 1900; rpt. in facsimile, Beijing: Zhonghua Shuju, 1983 [abbreviation *HSBZ*].

Wang, Zhongshu, *Han Civilization*, New Haven: Yale University Press, 1982.

Weld, Susan, "Covenant in Jin's Walled Cities: The Discoveries at Houma and Wenxian," Ph.D. dissertation, Harvard University, 1990.

Weber, Max, *The Protestant Ethic and the Spirit of Capitalism*, New York: Charles Scribner's Sons, 1958, rpt. 1976.

Weber, Max, *Economy and Society: An Outline of Interpretive Sociology*, ed. Buenther Roth and Claus Wittich, Berkeley: University of California Press, 1978.

Wheatley, Paul, *The Pivot of the Four Quarters: A Preliminary Enquiry into the Origins and Character of the Ancient Chinese City*, Chicago: University of Chicago Press, 1971.

Wu Hung, *Monumentality in Early Chinese Art and Architecture*, Stanford, California: Stanford University Press, 1995.

Wu Jiulong 吳九龍, *Yinqueshan Hanjian shiwen* 銀雀山漢簡釋文, Beijing: Wenwu chubanshe, 1985.

Xu Fuguan 徐復觀, *Zhongguo renxing lun shi* 中國人性論史, Taizhong: Donghai daxue, 1963.

Xu Jinxiong 許進雄, *Yin buci zhong wuzhong jisi de yanjiu* 殷卜辭中五種祭祀的研究, Taipei: Taiwan University, 1968.

Xu Xitai 徐錫台, *Zhouyuan jiaguwen zongshu* 周原甲骨文綜述, Xi'an: Sanqin chubanshe, 1987.

Xu Zhongshu 徐中舒, *Jiaguwen zidian* 甲骨文字典, Chengdu: Sichuan cishu chubanshe, 1989.

Yang Hongxun 楊鴻勛, "Xizhou Qiyi jianzhu yizhi chubu kaocha 西周歧邑建築遺址初步考察," *Wenwu* 文物 3 (1981), pp. 23–33.

Yang Kuan 楊寬, *Zhanguo shi* 戰國史, Shanghai: Shanghai remin chubanshe, 1980.

Yang Nansheng 楊南升, "Buci zhong suojian zhuhou dui Shang wangshi de chenshu guanxi 卜辭中所見諸侯對商王室的臣屬關系," in *Jiaguwen yu Shang Yin shi* 甲骨文與商殷史, ed. Hu Houxuan, Shanghai: Shanghai guji chubanshe, 1983, pp. 128–72.

Yantielun 鹽鐵論, *see* Huan Kuan 桓寬.

Yao Xiaosui 姚孝遂 and Xiao Ding 蕭丁, *Yinxu jiagu keci leizuan* 殷墟甲骨刻辭類纂, 3 vols., Beijing: Zhonghua Shuju, 1989.

Yates, Robin D. S., "Social Status in the Ch'in: Evidence from the Yun-meng Legal Documents. Part One: Commoners," *Harvard Journal of Asiatic Studies* 47.1 (1987), pp. 197–231.

Yates, Robin D. S., "New Light on Ancient Chinese Military Texts: The Development of Military Specialization," *T'oung pao* 74.4–5 (1988), pp. 212–48.

Yates, Robin D. S., "Body, Space, Time and Bureaucracy: Boundary Creation and Control Mechanisms in Early China," in *Boundaries in China*, ed. John Hay, London: Reaktion Books, 1994a, pp. 56–80.

Yates, Robin D. S., "The Yin-Yang Texts from Yinqueshan: An Introduction and

Partial Reconstruction, with Notes on Their Significance in Relation to Huang-Lao Daoism," *Early China* 19 (1994b), pp. 75–144.

Yates, Robin D. S., "State Control of Bureaucrats under the Qin: Techniques and Procedures," *Early China* 20 (1995), pp. 331–366.

Yinqueshan Hanmu zhujian zhengli xiaozu 銀雀山漢墓竹簡整理小組, *Sun Bin bingfa* 孫臏兵法, Beijing: Wenwu chubanshe, 1975a.

Yinqueshan Hanmu zhujia zhengli xiaozu 銀雀山漢墓竹簡整理小組, *Sunzi bingfa* 孫子兵法, Beijing: Wenwu chubanshe, 1975b.

Yu Yingshi 余英時, *Shi yu Zhongguo wenhua* 士與中國文化, Shanghai: Shanghai Renmin chubanshe, 1987.

Yunmeng Shuihudi Qinmu 雲夢睡虎地秦墓, Beijing: Wenwu Chubanshe, 1981.

Zhang Yachu 張亞初 and Liu Yu 劉雨, *Xi Zhou jinwen guanzhi yanjiu* 西周金文官制研究, Beijing: Zhonghua shuju, 1986.

Zhao Boxiong 趙伯雄, *Zhoudai guojia xingtai yanjiu* 周代國家型態研究, Changsha: Hunan jiaoyu chubanshe, 1990.

Zhong Bosheng 鍾柏生, *Yin Shang buci dili luncong* 殷商卜辭地理論叢, Taipei: Yiwen yinshuguan, 1989.

Zhongguo shehui kexueyuan Kaogu yanjiusuo 中國社會科學院考古研究所, ed., *Xiaotun nandi jiagu* 小屯南地甲骨, 2 vols. (vol. 1 in 2 parts, vol. 2 in 3 parts), Beijing: Zhonghua shuju, 1980–3.

Zhongguo shehui kexueyuan Kaogu yanjiusuo 中國社會科學院考古研究所, ed., *Xin Zhongguo de kaogu faxian he yanjiu* 新中國的考古發現和研究, Kaoguxue zhuankan, series A, no. 17, Beijing: Wenwu chubanshe, 1984.

Zhou Fagao 周法高, *Jinwen gulin* 金文詁林, Hong Kong: Chinese University Press, 1975.

Zhu Fenghan 朱鳳翰, *Shang Zhou jiazu xingtai yanjiu* 商周家族形態研究, Tianjin: Tianjin Guji chubanshe, 1990.

Zhu Fenghan 朱鳳翰, "Shang Zhou qingtongqi mingwen zhong de fuhe shiming 商周青銅器銘文中的複合氏名," *Nankai xuebao* 南開學報 3, 1983.

Zito, Angela, *Of Body and Brush: Grand Sacrifices as Text/Performance in Eighteenth-Century China*, Chicago: University of Chicago Press, 1997.

Zuozhuan 左傳; references are to Kong Yingda 孔穎達, *Chunqiu Zuozhuan zhengyi* 春秋左傳正義, in *Shisanjing zhushu* 十三經注疏, vol. 2, ed. Ruan Yuan 阮元, rpt. in facsimile, Beijing: Zhonghua shuju, 1980.

Index

References to figures, maps, and tables are in bold type. Chinese characters for names and terms are given on their first occurrence in the text.

www.ingramcontent.com/pod-product-compliance
Lightning Source LLC
LaVergne TN
LVHW090319210926
848480LV00001B/302
9780521027496